SMALL
IS
POSSIBLE

SMALL
IS
POSSIBLE

George McRobie

With a Foreword by Verena Schumacher

HARPER & ROW, PUBLISHERS

NEW YORK

Cambridge
Hagerstown
Philadelphia
San Francisco

1817

London
Mexico City
São Paulo
Sydney

Library of Congress Catalog Card Number: 79-2634
ISBN 0-06-013041-5   81 82 83 84 85 10 9 8 7 6 5 4 3 2 1
ISBN 0-06-090694-4 pbk   81 82 83 84 85 10 9 8 7 6 5 4 3 2 1

Many years of work on these matters have completely convinced me not only that small is beautiful but also that *small is possible* and has the future on its side.

E. F. Schumacher,
Sir Winston Scott Memorial Lecture,
Barbados, November 29th, 1976

CONTENTS

Part Three   What Small Makes Possible

FOREWORD

by Verena Schumacher

It was in 1955 that my husband first became interested in the problems of the developing countries. For six months he was seconded from his work at the National Coal Board, on the initiative of the United Nations, to be Economic Adviser to the Government of Burma – an experience that influenced him deeply. In fact, he described his response in the last speech he ever made. (It is reproduced in this book.)

The clash, not only of two technologies but also of two cultures, was vividly illustrated by a meeting he had with an American telecommunications engineer, who, exasperated by the lack of response of the Burmese people to his attempts to improve their country's communications, declared, 'The trouble with these people is that they're too goddamned happy!'

Also in Rangoon was a lady from New Zealand, who had been sent to inspect mental hospitals but was told, 'We haven't any. We don't lock people like that up in hospitals. You are not going to take them away from us.'

The temporary Economic Adviser was impressed, not only by the homogenous nature of the community, but also by its non-violent character. Clearly, such a people needed a non-violent form of technology for their material development – one which would not disturb their natural social cohesiveness, or clash with their religious beliefs.

Seven years later he undertook a similar mission to India, which confirmed his beliefs and produced results also described in this book.

When Fritz began to undertake such commissions to advise developing countries he felt a great sense of loneliness on his return. Most of his economist friends laughed at his low-cost, small-scale development ideas. Governments of the countries he advised paid only lip-service to his counsels. The industrialists, and most official development agencies, were interested only in selling the latest and most sophisticated hardware to the poor countries. But he found a friend who was willing to learn from him in George McRobie, who was then working with him at the National Coal Board. Their conviction that the problems of poor countries could not be met by the 'marvels' of Western technology led the two of them, along with Julia Porter, to set up the Intermediate Technology Development Group in 1965. Together they set themselves the task of helping people to help themselves.

So George McRobie is well qualified to write this account of what has happened so far.

This book shows how far and how quickly the movement has spread. Some of the organizations referred to existed before the I.T.D.G. was set up; others have no formal organizational links with it. But mostly their leaders say they were influenced by the ideas first widely promoted in *Small is Beautiful* – published only seven years ago.

There is now clear evidence that the need for more appropriate technologies is every bit as great in the materially richer countries. In 1975, for example, Fritz visited Prince Edward Island in Canada and found that although the main activity, agriculture, was tremendously advanced in its techniques and efficiency, many of the basic skills needed by the community had been lost: there was no one near at hand able to carry out simple building repairs and mend equipment. People had lost skills which used to be part of everyday life to their ancestors.

Unemployment is rising again in the highly industrialized countries, partly because of economic recession but also as a result of the success of technology in reducing the number of people needed to supply the demands of shrinking markets. Alternative technology practised in small groups can make a major contribution to the solution of the appalling social problems caused when human beings are denied the essential dignity of worthwhile work.

This book shows that a great deal of progress has been made, but there is enormous scope for more. It is fun to take part in an act of creation, as many people have found out. You can be creative without being an artist, but you can't be fully human without being creative. *Small is Possible* shows you that you are not alone in cherishing that most human ambition of becoming involved creatively in your work.

Fritz would have valued this book and wished success to the people whose work is described in it. Take courage, and start small.

ACKNOWLEDGMENTS

A book of this kind, which sets out to describe and in some measure interpret what is being done by thousands of people in many parts of the world, could not possibly have been written without a great deal of assistance.

To begin with, it draws on published and unpublished material of Fritz Schumacher's, and, for permission to use this, and for her encouragement, I am greatly indebted to his widow, Verena Schumacher.

For permission to reproduce in full Fritz's 1965 article, 'How to Help Them Help Themselves' (see p. 25), which was the first major public statement on intermediate technology, I am grateful to the Editor of the *Observer*.

Practically everyone at the Intermediate Technology Development Group at some point assisted in the book's preparation, especially Warren Adams, Steve Bonnist, Nicky Carter, Dennis Frost, Marion Porter, Mark Sinclair, Frank Solomon and David Wright. Several of the Group's associates and consultants gave me much of their time and expert knowledge, notably John Davis, Katherine Elliott, Peter Fraenkel, John Parry, Peter Stern and Stan Windass. The further reading lists were compiled by Marilyn Carr and John McRobie. I am particularly indebted to Charis Ryder for her professional editorial advice, and to Sarah Chedlow for compiling the index.

One of the pleasures of writing this book was the opportunity it gave me to renew old friendships and make new ones. In India I enjoyed the support and guidance of M. K. Garg, M. M. Hoda, Pandit Patankar and Anand Sarup, all of whom I first met in the mid-1960s, when we helped to found the first A.T. group to be started in India.

Among other friends in the developing world to whom I am indebted for material are Ghulam Kibria in Pakistan, A. T. Ariyaratne and Ton de Wilde in Sri Lanka, and Ben Ntim and John Powell of the Technology Consultancy Centre at Kumasi in Ghana. The account of the T.C.C.'s work owes much to the researches of Sally Holterman, who kindly allowed me to draw on her work.

E. M. Masale of the Department of Social Services, Government of Kenya, and Freddy Wood of the Bernard van Leer Foundation are among others who read drafts and made constructive suggestions.

On the other side of the Atlantic, it was my good fortune to have Bill Ellis of TRANET as my colleague during a memorable tour of A.T. centres in the U.S.A. Among many other American friends to whom I am grateful are Ann Becker, Peter Gillingham, Carol and David Guyer, Hazel Henderson, Ted Owens, John and Nancy Todd and Peter van Dresser; and in Canada, the Challenge for Change unit at the National Film Board of Canada, Bill Dyson of the Vanier Institute, Narasim Kothari in Sudbury, Jim McCrorie of the University of Regina, and Andy Wells and his team at Prince Edward Island.

My very special thanks I have reserved for last: to my son John, who contributed most of the section of the book on alternatives in Britain; to my son Peter for help with indexing; and to my wife Sybil who edited it all. Whether or not it is true that every book has behind it a great deal of support from friends and family, it is certainly true of this one.

1981   GEORGE McROBIE

SMALL
IS
POSSIBLE

E. F. SCHUMACHER: ON TECHNOLOGY FOR A DEMOCRATIC SOCIETY

The day before his death on September 3rd, 1977, Fritz Schumacher was speaking at an international conference at Caux, Switzerland. His theme was that the highly industrialized, not only the developing, countries of the world must start to devise technologies that are more in harmony with people, and with the environment, and less dependent on non-renewable resources. Isn't it time, he asked, that we started to put some real effort into building lifeboats in the form of technologies that are small, simple and non-violent?

The first thing when we think about what we call the Third World or the developing countries, or to put it more simply about the poor, the first thing we ought to realize, is that they are real. They are actual people, as real as you and I, except that they can do things which you and I can't do.

Mr McNamara, the President of the World Bank, published some statistics recently. Something in the order of 500 million people in this world, he said, have less than $50 a year. I don't know if there is a single person in this room who could survive on $50 a year. But they are surviving. They have a know-how that we don't have. They are real, and we must not think of them as poor little souls, and luckily we come along and we are going to develop them.

No, they are survival artists and it is quite certain that if there should be a real resources crisis, or a real ecological crisis, in this world, these people will survive. Whether you and I will survive, is much more doubtful. India will survive, though whether Bombay will survive is more doubtful. That New York will survive is an impossibility. Probably the same applies for London or Tokyo, and an awful number of other cities.

You cannot help a person if you yourself don't understand how that person manages to exist at all. This came to my attention quite a long time ago, a quarter of a century ago, when I first visited Burma and then India. I realized that they were able to do things that we could not do. But if I, as an economic magician, could double the income, the average income per head, in Burma from the £20 it was at that time to £40, if I could do that without destroying the secret pattern of life which enabled them to live, then I would have turned Burma, I am sure, into the nearest thing to paradise we know.

But coming from England I realized that we couldn't even survive on £200 per person. If I doubled the income of Burma from £20 to £40, whilst changing the pattern from the traditional Burmese pattern to the English pattern, then I would have turned Burma into the world's worst slum. Yet I listened to all the economists, and all the people who talked about development aid, and nobody ever talked about this particular factor, this pattern. They went into these countries like a bull into a china shop. They said, 'Step aside, now we will show you how to live properly when you are really rich.' The poor people of Burma stood on the sidelines and said, 'But we are not rich, so what you are showing us is of no interest to us, except the few people in our own countries who are already rich.'

So in Burma, more than twenty years ago, I concluded that overseas development aid really was a process where you collect money from the poor people in the rich countries, to give it to the rich people in the poor countries. Nobody intended this, but there was a blindness about this pattern of living which enables the poor to survive. And so we offered our goods which of course only people already rich and powerful could take.

Then I went to southern India. I was a lucky person because the right question occurred to my mind. Everything begins with a question, and the right question was, 'What sort of technology would be appropriate for rural India?' Surely not the technology of Pittsburgh, of Sheffield, or of Dortmund or of Tokyo.

Fate has given me the name of a shoemaker. If you want to be a good shoemaker, it is not good enough to make good shoes and to know all about making good shoes, you also have to know a lot about feet. Because the aim of the shoe is to fit the foot. But most of us never thought about this.

There used to be a story about a country that unduly indulged in central planning. They had developed the finest boot the world has ever seen and they ordered 500 million pairs of this boot, all of the same size. Well, that is what we tend to do, because we don't really think of the poor being real: we think that we have the answer.

When I had asked myself this question, 'What would be the appropriate technology for rural India or rural Latin America or maybe the city slums?' I came to a very simple provisional answer. That technology would indeed be really much more intelligent, efficient, scientific if you like, than the very low level technology employed there, which kept them very poor. But it should be very, very much simpler, very much cheaper, very much easier to maintain, than the highly sophisticated technology of the modern West. In other words it would be an *intermediate technology*, somewhere in between. And then I asked myself another question, 'Why do they not use an intermediate technology? Why do they not use boots that fit their feet?' And then I realized that intermediate technology was not to be found. I realized that in terms of available technology, either it was very very low or it was very very high; but *the middle had disappeared*. I therefore came to the conclusion that there was a tendency in technological development which I called 'the law of the disappearing middle'. (This only, I am sorry to say, applies to technology, so it is not a hopeful message for middle-aged gentlemen.)

You can verify this proposition if you go to a bookshop, where there you can get the latest publications and you can get the classics. But anything published in 1965 is out of print, it is unobtainable.

I was a farm labourer in northern England some thirty-five years ago. We farmed very efficiently a 300-acre farm with mainly animal-drawn equipment, which if you were able to buy it today would cost something of the order of $10,000. Not a single piece of that equipment is available today. It has disappeared. It has been replaced by, of course, far more sophisticated, powerful mechanized equipment, which would not cost $10,000 but $150,000.

And what does this mean? Oh, we say, this is progress. Yes, for some people it is progress. But it means that more and more people are excluded because maybe you and I could raise $10,000, once we had a farm, but the number who can raise $150,000 to equip the farm is very much lower.

Not so long ago I was invited to visit one of the most illustrious research and development institutions in England, where they develop textile machinery. The director showed me everything. These textile machines are so wonderful. They can do anything, at an unbelievable speed. So since there is a little boy in every man, I was utterly astonished and utterly delighted. And I asked, 'How much is such a machine?' I was told, 'Well, that machine would be £100,000.' And then I said to the director, 'It seems to me that you now can do everything.' He said, 'Yes, we can do everything now.' I said, 'Well then, why do you not stop?' So he said, 'Stop? Go against progress? Stop?' I said, 'This is a very expensive establishment. Why do we not say: Now, *basta*, enough.' So he said, 'I am surrounded by all these clever young people and they can still make an improvement here and a further refinement there. What is wrong with that?' And I said, 'Nothing is wrong with it except then this machine which now costs £100,000 will cost £150,000.' 'Are you against progress?' 'No,' I said, 'I am not against progress, but I am worried. Even now most of mankind is excluded. And when it is even more expensive, an even greater proportion is excluded.'

Our institutions are swarming with people who are wringing their hands about the overwhelming power of multinational companies. And at the same time applauding the technological development which makes production so complex and so colossally costly that only the multinational companies can carry it. This is the predicament not only of the developing countries but also of our own countries now. The middle way, which is also the democratic way that gives the little people some independence and what the young call 'doing one's own thing': that is being destroyed. And therefore we have throughout the world this atmosphere of tension, even of hatred.

There is little point in attacking the multinationals when the whole of society is bumbling along led by engineers and scientists who then introduce another complication, another speeding up.

Well, that is their job. But we as a society have not got enough philosophy or humanity to call a stop when a stop is indicated. Or at least to try and counterbalance it.

Now this fellow who took me through this resource establishment is a very thoughtful man and when I told him what I was worried about, he did then stop walking. And he said, 'But what can I do? If I go to the top man and say, "I think it is time to stop," he will say, "Yes, you have been looking a little bit jaded and you'd better take a holiday," and if I insist he will say, "I am grateful that you have given me notice so early. I already have in mind a very able successor." So I cannot stop.'

'Of course you cannot stop, but you can realize what you are doing, that this is not simply an interesting and productive and self-creating technological development, but it is a force that forms society, and forms it so that fewer and fewer people can be real people. At least you must be concerned with creating some counter-force to balance it up.'

So he said, 'And what would that be?' I said, 'Why do you not at least take ten per cent of your bright, innovating, creative engineers, and say to them, "Do not make this complicated machine even more complicated. Do not make this expensive machine even more expensive. What about the mass of mankind? They have got nothing. Take the simplest textile equipment and see whether you are not clever enough to make a much better job of it. To make it much more productive so that these people can make a decent living. This would be enough."'

I am afraid he has not done it. But this question has now come up throughout the world. The question is being asked, which I asked myself in another context twenty years ago in southern India, 'What is the appropriate technology to meet the very urgent problems that we are confronted with?' And people are coming to the conclusion: we do not have an appropriate technology for these problems.

What are these problems? Well, we do not have an appropriate technology for energy. Everybody agrees on that now. We are totally dependent on non-renewable sources of fuel, and that, of course, is really capital which we are consuming, it is not income. If we go along as before there will be an end to it.

We do not have a technology that is particularly kind to living nature around us, and so we have to take thought and concern

about an ecologically sound technology. That has now become
generally understood. But more questions are being raised. 'Do
we really have an appropriate technology from a human point of
view?' Well, I know many countries and I have talked to many
people who are exposed to our technology as factory workers and
so on. And I have had the experience that it is not very wise to ask
them if they enjoy their work, or if you ask them, do not wait for
the answer.

It is interesting to go back in history. Last year not only was the
bicentenary of the United States, it was also the bicentenary of the
appearance of a book by Adam Smith called *The Wealth of
Nations* – the basis of economics. Adam Smith said in effect, 'By
that which a person does all day long, he is formed. His work
forms him. And if you give him mindless work, he becomes a
mindless person. And he cannot be a good citizen, he cannot be a
good father in the family, or mother for that matter.' And then
surprisingly – or not surprisingly – Adam Smith goes on, 'But to
become totally reduced through mindless work is the fate of the
great majority of the people in all progressive countries.' He did
not say, 'This is terrible, they must not do it.' No, he had much the
same mentality: 'Well, that is just too bad, but that is the price
we have to pay.' And we all know that the human being has a
marvellous fortitude in tolerating the sufferings of others. And
even today we say to one another, 'This is not so bad. Of course
most of this work is not enjoyable, but it's got to be done, and they
actually enjoy it. Of course, I would not enjoy it.' Well, I was a
manual worker for quite a few years, but I did not suffer as much
as many of my friends are suffering, because I always could see the
light at the end of the tunnel. I always knew I would not have to do
it all my life. I can tell you, if I had thought that I would have to do
it all my life there would have been trouble.

No, we have not got an appropriate technology from a human
point of view. The sub-title of my book *Small is Beautiful* was
'Economics as if people mattered'. We do not approach economics
primarily from the point of view of people; we approach it from
the point of view of the production of goods, and the people as a
kind of afterthought. Of course, if they become redundant, well,
we have to pay them redundancy pay. If they have no opportunity
of using their skills, then we have to re-train them. If the work is so
noisy that they lose their hearing, well then we have to put some-

thing around their ears. They are factors of production. And this is the kind of industry we are now carrying into the so-called developing countries.

We are doing it at a time when in our heart of hearts we know that this kind of industry has no future. Nature cannot stand it, the resource endowment of the world cannot stand it, and the human being cannot stand it.

Already half of all the hospital beds in Britain and the United States are occupied by people whose problems are mental not physical. This kind of industry has no future. If the society that claims to be the biggest and the richest and has an income per head twice that of Britain, namely the United States of America, if they have not solved the problems of harmony, of the city, of poverty, they are not on the right road. So what shall we do?

When we begin to suspect that we are not on the right road, then of course we get a lot of fanatics. And a fanatic is a person who, when he senses that he is doing the wrong thing, redoubles his efforts. We have plenty of those. I call them the people of the forward stampede. They have a slogan, and blazoned on their banner is 'a break-through a day keeps the crisis away'. They are stampeding us into greater and greater violence.

But now there is another great ground-swell of people whom I call the 'home-comers', who say, 'Well really, the purpose of our existence on this earth cannot be to destroy it. The purpose of our existence can't be to work ourselves silly and to end up in a lunatic asylum. Let's reconsider it.'

I want to make the point that these people exist, in my experience, in all societies, and the people of the forward stampede also exist in all societies. I was recently on the other side of the Iron Curtain, where they explained to me at great length that their system was so much better than our system. And finally they said, 'In any case, the Western economies are like an express train hurtling at ever increasing speed towards an abyss.' Then there was a short pause, and they added, 'But we shall overtake you.'

That is the automatism of 'progress'. That is the rivalry. It is a kind of fraud. And so it is necessary for us to step back and have a new look, and at least to create some sort of counterweight.

Initially, with regard to the Third World, the so-called developing countries, some friends of mine said, 'Well, let's do something about this', and we set up an organization which we called the

Intermediate Technology Development Group Ltd, not to kill off the high technology, because we couldn't do that anyhow, but to fill this gap, this middle that has disappeared. And perhaps thereby to make it possible to overcome the fateful polarization which modern technology has produced, under which the rich become richer and the poor become more desperate, and society disintegrates – something that you can observe on a world scale, and you can also observe in all big countries. Small countries are more beautiful.

The polarization is now so great that you can keep the consequences plastered over only with enormous welfare expenditure. Welfare will keep people afloat, but does not integrate them in society. And in the United States, for example, you have many people who are third generation welfare recipients.

And even the great United States has come to the conclusion that with the present easily available technology, we cannot solve the problem. So they have set up a national centre for appropriate technology, not for the developing countries, but for the United States. They said, 'We must re-think technology, and try to make it appropriate to our actual problems.' These problems are not simply more and more production. The actual problems are the reintegration of a sizable proportion of the total population into the mainstream of society.

Similar things are happening in all advanced countries. So now we are in the position of talking about appropriate or intermediate technology in a much more convincing way. When people in the Third World say to me, 'Well, if it is such a good thing, why don't you do it?' I say, 'Yes, we do do it.'

What is this new look at technology? We have to ask the right questions. Is it relevant to the real problems we have? I will give an example. Some of my French friends may be hurt by it, and our English friends, but nobody else. Let's ask the question about the appropriateness of technology in connection with Concorde.

The proper question to ask is, 'Is it a very intelligent development in terms of the energy situation of the world?' You have to give the answer, I don't. Because that is a big problem. Is it a good thing in terms of environmental quality? It may be, of course; people may say the environment is greatly improved by the sonic boom. Is it appropriate technology in terms of fighting world poverty? Does it help the poor? Is it an appropriate technology from a democratic point of view? Perhaps getting a greater equality among people?

You can take every single problem of this society, and you have to ask, is the technological development appropriate? Or is it some sort of little boy's engineer dream? We can do it, so let's do it.

These questions now have to be faced, and I am glad to say in more and more countries there are now groups who normally sail under the name of something like 'technology assessment'. They are calling companies to order and asking: 'This development, you have assessed it in terms of power or glory or profitability, but that is not good enough. So let's ask, is it relevant to the real problems of mankind?'

We could go on for a long, long time talking about this. I just give you the outcome of our work in terms of ideas when we set up this organization twelve years ago. It was created out of nothing. There is no money behind it. Just a few ideas. We have now in Britain something of the order of a thousand people, highly qualified people, working in various parts of society with us on the creation of an appropriate technology.

You have to go to a bit further than the word intermediate. We gradually came to four criteria. What is lacking is a small-scale technology, we know how to do things on a big scale. But if we only know how to do things on a big scale for a big market, then our industry will go into the big cities and we have, as we can observe all over the world, another polarization – vast congestion in a few places and enormous emptiness elsewhere. Again, you only have to look across the Atlantic at countries represented here like Canada or the United States, and you find precisely this polarization.

Many communities in North America are coming to the Intermediate Technology Group and saying, 'We have become colonies of the big metropolitan areas! We don't want to be colonies, we want to be ourselves. We want to have our own society, not simply provide raw materials for Chicago. We want to have jobs in Montana. We are like a Third World country in Montana. That is a fact. Out of a hundred graduates from our universities, eighty-five cannot find a job in Montana, they have to go to the big cities. Why can't we have jobs in Montana? Because we haven't got the appropriate technology. The big technology only fits into the big concentrations of populations. Are we really so limited that we can't create an appropriate small-scale technology?' That is question number one.

Question number two. Of course you have to go to a big city if your processes of production are so complex that you need the highest experts by the hundreds. You can't find them in Montana or in Regina. So you go to Toronto, you go to Chicago. So can't we create a technology which is not so complex? It takes better engineers – even a third rate engineer can make a complicated thing even more complicated. It takes a bit of genius to recapture the basics and simplicity.

The highly complex technology does not fit into the rural areas of the world and so this polarization will go on unless we make it our business to create such a technology.

Number three is what I have already talked about: the cost per workplace. The costliness of capital equipment has been sky-rocketing, so, as I said before, there are increasingly only the multinational companies left who can afford to create workplaces, and the little people are left out. And if there are unemployed, it is just too bad. Can't we make it our business to create a technology which is cheap per worker? This requires entirely new thinking.

And the fourth point is slightly different. I know I may arouse some opposition when I say this. Modern technology has become increasingly violent. Violence is not just a matter of one person hitting another person over the head, it is employing violent means. We have this in agriculture, where we scatter around very violent chemicals, we call them pesticides, which means killer substances. On this thin living film of the earth on which all life depends we are scattering millions of tons of killer substances. Whatever you may think, it is a violent technology. And the spirit of violent technology has invaded the medical profession. We want quick results by violent means. So much so that we claim that a very high percentage of all illnesses are induced by the doctor. The only advice we can give anybody is to avoid the doctor when you are ill.

Of course the greatest readiness to resort to violence we are now experiencing in our attempts to cope with the energy problem, where we are prepared to put into the world large amounts of plutonium, a substance of a really unbelievable ghastliness, which the good Lord never made. He knew where to stop. It is a man-made thing. And it will be there for all time: the half-life radioactivity for plutonium is 24,400 years. In fact, before it is really quite harmless it takes a matter of 3 million years.

All this is a readiness to apply extremely violent processes to that sacred and unbelievably complex system called nature. We don't know what we are doing.

Of course we have wonderful scientists, who give us the assurance that all is well, don't worry. This is the blind leading the blind with a vengeance. It is not necessary to be violent. We know that in agriculture, in medicine, in energy, in any other subject you may care to think of, there are people who are very often called, or used to be called cranks, who know how to produce enough food, how to keep healthy, without any violent methods. All this is possible, but it hasn't had any support at all from governments, and very little either from academia, or business.

Governments everywhere put a great deal of money into agriculture, which goes into chemicals and mechanization, but the organic farmer who is showing what is possible without chemicals, he gets no help. And when it comes to most academics, they simply get angry when they are told that there are methods that are more elegant than the violence of their chemicals.

These are the four criteria that have crystallized out of this work. We don't feel we are unsuccessful, although we don't represent anything, we don't have power, we have little money. We have had a certain influence because we are happy to work with everybody or anybody. I am often challenged and asked, 'Do you work with academia?' I say, 'No, not with academia!' 'Do you work with business?' 'I don't work with business.' 'Do you work with governments?' 'No, no! You can't work with them because they are all committed to that monster technology.' 'Well, whom do you work with?' I work with *people* from business, academia and government. In the vast and seemingly monolithic structures we have individuals who come to us and say, 'What you are doing is interesting.' And they are carrying it into their companies and into government departments, and even into the universities.

I would say that part of the job we have is to try and persuade people that they should give some thought, some systematic thought to it. In this connection we say, 'Look, even the most wonderfully designed ocean steamer carries lifeboats, not because some statistician has predicted that the steamer will run into an iceberg, but because icebergs have occasionally been seen. Isn't it

time that the modern world provided some lifeboats?' Of course
you don't put all your money into the lifeboat, you don't put all
your research and development into the exploration of small,
simple and non-violent technology. You have to go on making a
living, but 5 per cent or 10 per cent could be so spent.

Some businesses are doing it. And if a big business comes and
says, 'I will give this thinking a chance,' they have never felt sorry.
They suddenly realized that really the construction of the universe
is far more benign than they ever thought. You don't have to be so
violent. We are now quite intelligent enough to create appropriate
technologies, if we really think before we act, and think in these
wider terms.

In order to do anything, we find that it is necessary, as I said
before, to take a very co-operative attitude in the various panels
and working parties we have set up in the Intermediate Technology Development Group. We try to achieve in every one of
these organizations what we call the ABCD combination.

That is just so that you should remember it more easily. 'A'
stands for the administration, people from government. Let's have
some of them on the working group, as persons, not representing
governments. They know how to pull the strings, and they control
a lot of money; they are the tax-gatherers and spenders. That is the
'A' factor, administrators.

'B' stands for business. Now the business intelligence is the
intelligence, the discipline, to make things viable so that they can
survive. To create a thing that cannot survive is a waste of time.
We need this intelligence.

The 'C' factor are the communicators. The people of the word,
research people, people who have got time to think and to write.
They solve problems, and there are plenty to be solved. But never
let them act alone, because they are playful little souls. They like a
problem, whether it be a chess problem or a problem that means
something, and when they have solved it they mark it 'top secret'
and file it away somewhere and turn to the next problem. But if
the 'B' factor is sitting next to them, the business man, he says,
'We spent money on this. We must now bring it out and make it
viable.' So this is a very healthy combination.

And 'D' are the democratic organizations of society, they are
the labour union people, the women's organizations, the ecological people; happily every country is full of them. You don't

want to do this in an élitist way, and wherever you succeed in getting ABCD together, as persons, they have a good time. They really enjoy it, because initially they have a very low opinion of each other, very low. And then they realize that they are actually all quite intelligent and useful people.

Wherever we succeeded in getting the ABCD combination we found that things became possible that everybody had thought were quite impossible.

All I can say is, the whole thing doesn't cost a great deal of money, and people who join it enjoy it.

It was in the summer of 1977 that Fritz Schumacher decided that there should be a follow-up to *Small is Beautiful*, and that it should be entitled *Small is Possible*. The title describes its purpose. It was to be factual and informative about who was doing what, where, to carry into practice the ideas expressed in *Small is Beautiful*. The aim of the book, as he saw it, was not simply to strengthen the existing alternatives network that had grown up in developing countries and on both sides of the Atlantic, though it should certainly serve that purpose (many appropriate technology groups have very little knowledge about others working in the same field). Its chief function should be that of getting more work done to develop and introduce appropriate – small, simple, capital-saving, non-violent – technologies and their supporting institutions. 'An ounce of practice is worth a ton of theory' was one of Schumacher's favourite maxims: show, to as many people as possible, what some of them are already doing, and that – far more than any theoretical argument – will enlist their understanding and their support. The point of his argument was that most books on the practical aspects of development are written by experts chiefly for other experts. It is not to deny the value of that kind of activity – which is the conventional way that ideas and theories are arrived at – to assert that it is sometimes more important to make something work than to wait for the experts to give their judgment on whether it is possible. It is, after all, experts who tell us that, aerodynamically speaking, the bumble bee is incapable of flight.

I have tried to follow these guidelines in this book. It is not an assembly of economic or technical case studies; it describes rather than analyses the remarkable initiative and skill that tens of thousands of people are displaying in the process of making small technologies work for them, for others, and ultimately for all of us.

Many people associate the concept of intermediate technology with the developing countries and it was, of course, in that context that Schumacher first put forward the idea. Technologies that are small, simple and capital-saving are more appropriate to the needs and resources of poor countries than are the large-scale, labour-saving technologies developed in the West during their heyday of cheap energy. It was not long after the Intermediate Technology Group was started that it began to become evident that the rich countries, too, needed a new kind of technology, appropriate to the conditions in which they now find themselves. That these are crisis conditions there is no room for doubt – a crisis compounded of oil shortages, environmental destruction and a human revolt against depersonalized work and growing unemployment. A rapidly growing number of individuals and groups on both sides of the Atlantic have concluded that the only sane alternative lies in the direction of technologies that are relatively small, simple, capital-saving and non-violent, and economics as if people mattered.

That Schumacher was firmly of this opinion is revealed in what he said and wrote after *Small is Beautiful* had appeared, and some of his best essays on this theme have recently been published in a collection entitled *Good Work*. During my own conversations with him shortly before he died in September 1977, he had become increasingly convinced that the need of the rich countries to restructure their technologies and life-support systems was, if anything, more urgent than that of the poor countries.

These ideas are also reflected in this book. It starts with Schumacher's last talk, on appropriate technology for the industrialized countries. It goes on to describe what is being done by the Intermediate Technology Development Group and its counterpart organizations overseas, including those in Africa, India and Latin America; and also what is being done by the alternative technology movements in Britain, the U.S.A. and Canada.

These are the groups and organizations which have been putting Schumacher's ideas into practice during the past decade. They are showing that in almost every branch of human activity, in rich and poor countries alike, it is possible to create lifestyles and technologies on a human scale which are low-cost, sparing in their use of resources, non-violent towards nature, and, therefore, sustainable.

Part One

INTERMEDIATE TECHNOLOGY:
An Idea in Action

1

THE FORMATION OF THE INTERMEDIATE TECHNOLOGY DEVELOPMENT GROUP

The critical role of technology in economic development, and especially the importance of technological choice, was first brought into focus by Fritz Schumacher in 1962. In a report prepared for the Indian Planning Commission at the invitation of the then Prime Minister, Pandit Nehru, he formulated the concept and gave it a name:

> It requires no lengthy argument to agree that India is 'long' in labour and 'short' in capital. This means that she requires a level of technology, or 'capital investment per workplace', that is likely to be very different from that current in the Western countries, which are 'long' in capital and 'short' in labour. At present, in India as in all other developing countries, the most primitive exists side by side with the most advanced – an artisan employing five rupees' worth of tools, and workers minding machines worth fifty thousand rupees. But the intermediate industrial technology which would really suit India's conditions does not exist in an articulated form, except perhaps accidentally . . . If, therefore, it is intended to create millions of jobs in industry, and not just a few hundred thousands, a technology must be evolved which is cheap enough to be accessible to a larger sector of the community than the very rich and can be applied on a mass scale without making altogether excessive demands on the savings and foreign exchange resources of the country . . .
>
> I believe that it is not very difficult, from an engineering point

of view, to devise such a technology, provided that the engineers can be told in fairly precise terms what is wanted. It is no use simply asking for 'an intermediate technology', nor does it suffice to say that we want a technology that employs the minimum of capital, for the minimum is zero. If one specified 'the minimum of capital consistent with profitable production at a wage of at least two rupees a day', one would, I think, be asking the engineer to go beyond his competence and become a businessman. The simpler the specification, the better will be the results, and the simplest would be a statement of the amount of money to be spent on each workplace.[1]

Schumacher went on to suggest that, as a first approximation, the average investment per workplace in manufacturing units, suitable for widespread rural industrialization, should be of the order of Rs 1,000, excluding the cost of building. Detailed design studies should be undertaken on this basis for all industries envisaged for the rural areas. These studies would themselves disclose whether the average of Rs 1,000 was realistic, and what variations from the average might be required by different industries.

This was the first concrete expression of the concept of intermediate technology, an idea that within the next decade was to transform the thoughts and actions of a generation, both in the poor and the rich countries of the world.

What was finally crystallized in the report to the Indian Planning Commission had, in fact, a long history in Fritz Schumacher's mind. Like several of Schumacher's best ideas, its origins lie partly in his wartime experience as an agricultural labourer, which endowed him with an abiding love of the soil and an understanding of the need to treat it non-violently as a living substance. In part it stems from his postwar work in a war-shattered Germany when, as he often said, he first realized the meaning of the phrase that knowledge is power – the knowledge and the skill that lies in the minds and the hands of ordinary people and which is the source of all wealth. Earlier, too, his work with Beveridge on full employment policies had developed his awareness of the human consequences of large-scale industrialization. It was with these experiences fresh in his mind that he took on his first assignment in a developing country, as economic adviser to the government of Burma, in 1955.

In that year he wrote a short paper, 'Economics in a Buddhist Country',[2] later to be elaborated as a chapter in his first book, *Small is Beautiful*.[3] This was, to my knowledge, his first critique of the impact of Western economics on the life and culture of people in developing countries. It was, to be more precise, both a critique of the notion of limitless and completely indiscriminate growth, and the inability to distinguish between renewable and non-renewable products, on which Western economic development was founded, and it was also a warning of the destructive effect that this kind of economics could have on the developing countries.

This and a companion paper, 'Non-Violent Economics',[4] attracted the attention of Prime Minister Nehru and some of his advisers in India, and in 1961 Schumacher was invited to be a key speaker at an international seminar, Paths to Economic Growth, held in Poona. 'How', he asked, 'can the impact of the West be canalized in such a way that it does not continue to throw the people into apathy and paralysis? . . . We should be talking about getting the people to use their own labour power with their own intelligence (which is not incapable of picking up improved methods from outsiders), and their own local resources and materials to provide, in the first place, for their own fundamental needs, which are food, clothing and shelter, and certain communal assets like roads, wells and communal buildings.'

Six months later, in a paper written for the Gandhian Institute of Studies, entitled 'Levels of Technology',[5] he had virtually arrived at the concept of intermediate technology, when he introduced the idea of 'cost per workplace'. Speaking of a scene known to all developing countries, namely, a modern factory employing relatively few workers and surrounded by masses of unemployed people, he said:

> What is the chance of providing modern factory employment for all or most of these people who are hanging around the factory gates, hoping against hope? A very simple calculation may help here. Let us look at the national income statistics of an advanced country employing the level of technology which is represented by the factory in question, say, the United Kingdom. Annual capital consumption per head of the working population, as covered by depreciation allowances, amounts to

roughly £100. From this we can deduce, without any pretension to statistical accuracy, that the average capital cost per workplace would be something of the order of £1,000. Now this sum of £1,000 is roughly equal to annual average (national) income per head of the working population. In other words the creation of one working place and what goes with it, at the average level of technology currently represented by the British economy, would require one year's income of one man. Since this one year's income, at £1,000, is rather high, Britain can 'afford' a fairly high level of technology, that is to say, a high capital investment for each workplace and everything that goes with it. The converse, of course, is equally true: the annual income is as high as it is, largely because of the great amount of capital equipment that has gone into each workplace.

But we shall not see the decisive point if we focus our attention on the second statement, true as it is. For we are starting from a position when incomes are low and there is very little availability of capital; to dwell on the situation that would exist if great wealth had already been attained would merely deflect attention from reality.

It would seem to be obvious, therefore, that a poor country such as India, for example, could not afford a level of technology that required the investment of £1,000 per workplace, representing not one, but perhaps twenty or thirty years' income of one (average) man. That is to say, even if certain small 'islands' of such a level of technology were created (as they have been) in various parts of the country, there would seem to be no hope that this level of technology would generally establish itself throughout the country within a reasonable period of time.

When, some eighteen months later, he presented his proposals to develop 'intermediate technology' for India's rural areas, the idea was very coolly received by the majority of the Indian Planning Commission. The Rural Industries Section of the Commission were enthusiastic and tried to get some action, but they were very much in the minority. Several senior members of the Commission sponsored a conference[6] in Hyderabad early in 1964 in the hope of launching a nationwide programme of research and development on intermediate technologies, but at that time neither the government nor the Planning Commission would entertain the idea.

Meantime, in Britain, the Overseas Development Institute had published Schumacher's report to the Planning Commission in 1963, and in the following year he presented a more precise and elaborate statement of his arguments to the 1964 Cambridge Conference on Rural Development, an event which brought together many of the world's leading development economists. This paper, 'Industrialization through Intermediate Technology'[7] argued that:

> Industrial concentrations based on the latest technology may give you the best output for capital invested and the best rate of economic growth; but the political and human costs in unemployment in the countryside, in resentment against a policy that leaves eighty per cent of the population in rural areas worse off than they were before, are so great that society may disintegrate and the economy run down.

The support being given to concentrations of large-scale industry, Schumacher continued, could only exacerbate the twin problems looming over virtually every developing country; mass rural unemployment and under-employment, and mass migration to the cities. Moreover economic growth, a purely quantitative concept without any qualitative determination, cannot be accepted as a rational objective of policy. Accordingly he argued that the most urgent task of the industrialized countries was to help poor countries to develop and employ intermediate technologies that use less capital per workplace; and to adopt a strategy of rural industrialization based on small-scale manufacturing using intermediate technologies, with production largely from local materials for local use.

To say that these arguments struck the distinguished academics at the Cambridge conference as not only controversial but downright heretical would be an understatement. It sparked off a row that enraged Schumacher's critics and delighted his friends, for it was generally acknowledged that he got the best of the exchanges. Within a few years even the most diehard of the conventional economists had to admit that he had been right. Like that of many a prophet before him, Schumacher's vision was at first denied by his contemporaries, by the economists who were professionally best equipped to understand its full implications. By 1968,

> . . . experience had redressed the balance of the argument on Dr Schumacher's side. The best of plans based on maximum productive efficiency may yet make bad economic policy. At the end of the decade the divergence of the means of relieving unemployment from the means of maximizing economic growth became inescapable.[8]

But that was four years later. At the time the majority opinion, both at the Cambridge conference and among development economists and planners generally, was anti-intermediate technology. In mitigation, it should be remembered that those were heady days for economists. The shaky premises on which they stood had not yet been revealed. Grandiose economic development plans were all the rage abroad, and at home economists headed the onward stampede, as Schumacher called it, in pursuit of limitless economic growth based on an inexhaustible supply of cheap oil.

The fact remained that the idea of intermediate technology had not been at all well received. It had been virtually rejected in India, and required too much of a volte-face by most of the leading development economists and planners for it to be acceptable elsewhere. There was a danger that it would remain merely an interesting subject for academic discussion, to be talked to death. That was when we decided to form an action group to do something about it ourselves. ('We' comprised Schumacher, Julia Porter, who was running the Africa Development Trust, and myself.)

The genesis of the Intermediate Technology Development Group was in May 1965, when some twenty of us who were sympathetic to putting the idea into practice met at the offices of the Overseas Development Institute, then in Piccadilly, London.

The starting of the group was very much an act of faith. In Schumacher's words,

> We had no money, there were just a couple of friends of mine, like myself professional people with full-time work and families to support. But when you feel that something is necessary, you

can't simply go on talking about it – you have to talk for a certain while, but then the moment comes – I tell you, a frightening moment – when you have to take the existential jump from talking to doing, even if you have no money.[9]

No money at all was exactly what we had for the first few months; and therefore no real way of establishing contact with the outside world. We did not even know with any certainty whether we would be supported by people working in developing countries, or indeed how long we could keep the working group going.

This problem was solved for the Group in a dramatic and unexpected way. At the end of August 1965, the *Observer* published, in their Weekend Review, an article they had commissioned Schumacher to write several months earlier but had not used until then. It is reproduced in full here because it is his first statement on the subject that is addressed not to economists but to the public at large – and also because it launched the Group.

How to Help Them Help Themselves

The Western world spends hundreds of millions of pounds on aid to developing countries. But what if this aid, so far from reducing misery, is actually increasing it? This is the startling view which a distinguished economist puts forward here, together with the outline of a radical new approach to the whole problem.

Mass unemployment (as distinct from the limited unemployment which might arise out of Britain's present economic crisis) has been unknown for almost 30 years, ever since the 'Keynesian revolution' – in most of the *developed countries*, that is to say. It is quite different in large parts of the 'developing' or underdeveloped world.

'Developing countries,' says a recent study in the *International Labour Review*, 'cannot include the goal of full employment among their immediate planning targets.' The Third Five-Year Plan in India showed higher unemployment at the end of the period than at the beginning. Much the same is true of Turkey and, in fact, of most of the larger developing countries.

In short, mass unemployment in these countries is being accepted as inevitable and unconquerable, even as 'necessary for sound growth', in much the same way as was the case in the advanced countries before Keynes. The arguments, to be sure, are somewhat different. The developing countries, it is said, cannot have jobs for all because they are short of capital.

Unemployment and under-employment in developing countries are most acute in the areas outside a few metropolitan cities; so there is mass migration into these cities in a desperate search for a livelihood: and the cities themselves, in spite of 'rapid economic growth', become infested with ever-growing multitudes of destitute people. Any visitor who has ventured outside the opulent districts of these cities has seen their shanty towns and misery belts, which are often growing 10 times as fast as the cities themselves.

Current forecasts of the growth of metropolitan areas in India, and many other developing countries, conjure up a picture of towns with 20, 40 and even 60 million people – a prospect of 'immizeration' for a rootless and jobless mass of humanity that beggars the imagination.

No amount of brave statistics of national income growth can hide the fact that all too many developing countries are suffering from the twin disease of growing unemployment and mushrooming metropolitan slums, which is placing their social and political fabric under an intolerable strain.

The suspicion has been voiced (and cannot be dismissed out of hand) that foreign aid, *as currently practised*, may actually be intensifying this twin disease instead of mitigating it; that the heedless rush into modernization extinguished old jobs faster than it can create new ones; and that all the apparent increases in national income are eaten up, or even more than eaten up, by the crushing economic burdens produced by excessive urban growth. It is rather obvious that a man's cost of subsistence – something very different from his standard of life – rises significantly the moment he moves from a small town or rural area into 'megalopolis'.

Right road

No wonder, then, that there is a widespread search for a new approach. If the current methods and types of foreign aid produce such questionable results, it is perhaps a bit superficial merely to demand an increase in the volume of aid. Let there be such an increase by all means, but only when we are sure that we are on the right road. How can *we* judge whether we are on the right road or not?

Central planners take as their decisive criterion the rate of growth of Gross National Product, that is, the developing country's aggregate of money incomes. This is highly misleading. If the rise in G.N.P. is accompanied by rising unemployment and an increase in social tensions, the outcome of the enterprise is unlikely to be satisfactory.

The first task of any society is surely to avoid the extremes of misery and frustration. If 'the people' are left out of development planning; if economic growth merely intensifies, as it tends to do, the appalling features of the 'dual economy' – a small sector of opulence surrounded by an ocean of misery; then the final outcome will be disastrous.

The primary task of developing countries now afflicted by mass unemployment and mass migration into a few metropolitan areas would therefore seem to be clear: to go straight into battle with these evils. This means:

1. Workplaces have to be created in the areas where the people are living now, and not primarily in metropolitan areas into which they tend to migrate;
2. These workplaces must be, on average, cheap enough so that they can be created in large numbers without this calling for an unattainable level of savings and imports;
3. The production methods employed must be relatively simple, so that the demands for high skills are minimized, not only in the production process itself but also in matters of organization, raw material supply, financing, marketing, and so forth;
4. Production should be largely from local materials for local use.

These needs can be met only:

A. If there is a 'regional approach' to development;
B. If there is a conscious effort to develop what might be called an 'intermediate technology'.

A given political unit is not necessarily of the right size as a unit for economic development. If vast and expensive population movements are to be avoided, each 'district' with a substantial population needs its own development. To take a familiar example, Sicily does not develop merely because Italian industry, concentrated mainly in the north of the country, is achieving high rates of economic growth. On the contrary, the developments in the north of Italy tend to increase the problem of Sicily through their very success, by driving Sicilian production out of existence and draining all talented and enterprising men out of the island.

If no conscious efforts are made to counteract these tendencies in some way, success in the north spells ruination in the south, with the result that mass unemployment in Sicily forces the population into mass migration. Similar examples could be quoted from all over the world. Special cases apart, any 'district' within a country, if it is being bypassed by 'development', will inevitably fall into mass unemployment, which will sooner or later drive the people out.

Each 'district', ideally speaking, would have some sort of inner cohesion and identity, and would possess at least one town to serve as the district centre. While every village would have a primary school, there would be a few small towns with secondary schools, and the district centre would be big enough to carry an institution of higher learning.

This need for internal 'structures' is, of course, particularly urgent in large countries, such as India. Unless every district of India is made the object of development efforts, so to say, for its own sake and in its own right, all development will concentrate in a few places – with devastating results for the country as a whole.

It is obvious that this regional or district approach has no chance of success unless it is based on the employment of a suitable technology. Here we come to the crux of the matter.

Western technology has been devised primarily for the purpose of saving labour; it could hardly be appropriate for districts or regions troubled with a large labour surplus. Technology in Western countries has grown up over several generations along with a vast array of supporting services, like modern transport, accountancy, marketing, and so forth: it could hardly be appropriate for districts or regions lacking these paraphernalia.

This technology, therefore, 'fits' only into those sectors which are already fairly modernized, and that means – some special cases apart – the metropolitan areas, comprising, say, 15 to 20 per cent of the whole population.

What, then, is to become of the other 80 to 85 per cent? Simply to assume that the 'modern' sectors or localities will grow until they account for the whole is utterly unrealistic, because the 80 per cent cannot simply 'hold their breath' and wait: they will migrate in their millions and thereby create chaos even in the 'modern' sectors.

Much cheaper

The task is to establish a tolerable basis of existence for the 80 per cent by means of an 'intermediate technology' which would be vastly superior in productivity to their traditional technology (in its present state of decay) while at the same time being vastly cheaper and simpler than the highly sophisticated and enormously capital-intensive technology of the West.

As a general guide it may be said that this 'intermediate technology' might be on the level of £70–£100 equipment cost per average workplace. At this level it would be cheap enough to be 'within reach' of the saving efforts of the more enterprising minority in these areas; it would be simple enough to catch on with them; its functions would depend neither on the regular supply of highly refined raw materials nor on near-perfect systems of organization; in short, it would simply 'fit' into the social context as a whole without depending on the availability of factors which, as experience shows, cannot be depended upon.

Some economists have argued, first, that such an intermediate technology would be a waste of scarce capital resources – that it would yield less output per unit of capital than could be obtained from the most highly capital-intensive mode of production; and, second, that the products of such a technology could never be competitive.

Both arguments deserve consideration. Countries short of capital would be foolish to squander their scarce capital resources on relatively unproductive projects and certainly must look for the highest possible output-capital ratio. But the question of what level of technology in fact produces the most favourable ratio is not a question of economic theory but of applied engineering.

Dogmatic pronouncements on this point are worthless: let us have design studies, and we shall see.

This applies to the second objection. Labour-intensive methods of production may or may not produce goods at competitive prices; there are no laws of nature or man to decide the question in the abstract.

In this matter an ounce of practice is worth a ton of theory. What is wanted is nothing more nor less than a series of humble design studies. Let us see what can be done by relatively simple means, with mainly local materials, local labour, and low-cost capital equipment – equipment which would be simple enough also to be made locally.

Basic goods

Design studies undertaken in India have demonstrated that many products are suitable for 'intermediate technology' production – practically all basic consumers' goods, building materials, agricultural implements, and many kinds of equipment for the 'intermediate technology' industries themselves – and that these products can be fully competitive with those of Western technology.

Every industrialist knows, of course, that there is still a long and arduous road to travel between a design study and its practical implementation. 'Intermediate technology' is no magic wand. Countries like India need millions of additional workplaces, which modern technology cannot possibly provide within the foreseeable future. To provide them by means of an 'intermediate technology' would be possible at any rate within the given and known limits of internal capital formation *plus* foreign aid. It would, to be sure, require a great organizing effort on the part of very many people.

Would such an effort be forthcoming? No one can say it would not. The lack of entrepreneurial ability, which the central planners in developing countries so frequently deplore, is itself largely the result of their present fixation on the naive idea that what is best for the rich must also be best for the poor. Most of the ambitious planning currently undertaken leaves 'the people' helpless and disheartened: it does not 'fit' into their way of life and remains outside their power of self-help.

This is not to say that all projects on the level of Western technology are useless. Let them continue, although perhaps with rather more caution than has been shown in the past. What appears to be certain beyond doubt is that they will have to be supplemented by a determined effort to reintegrate the jobless millions into the economic process by means of something along the lines of 'intermediate technology'.

Main needs

Professor Gadgil, the doyen of Indian economists, has asked that 'the process of evolving and adopting intermediate technology . . . should be the centre of interest of the plan of industrialization of the country.' Foreign aid will be fruitful, instead of destructive, only if it recognizes these paramount needs and makes Western intellectual resources available to meet them.

This is not a long, expensive or particularly difficult task. Fragments of this 'intermediate technology' exist all over the world, in the advanced countries no less than in the poverty-stricken ones. A new approach is needed, a systematic effort to collect them and develop them into practical blueprints for industrial action. To quote Professor Gadgil again, this 'should claim the attention of the ablest scientists and technicians in the country.'

What stands in the way? Perhaps a kind of technological snobbishness which regards with disdain anything less than ultra-modern? Perhaps a certain callousness in the attitudes of privileged minorities towards the immense suffering of their homeless, jobless, miserable fellow-men? Or is it lack of imagination on the part of the planners in resplendent offices who find ratios and coefficients more significant than people?

Whatever it is, millions of people in the wealthier countries are today moved by a genuine desire to help those who live in misery, and the elemental force should be capable of overcoming all petty preoccupations. 'Intermediate technology' can help the helpless to help themselves.[10]

After this we were left in no doubt about whether we were on the right track. Letters of support and encouragement poured in from all over the world. Many more people in Britain offered to help, and a volunteer panel of scientists and engineers was formed to help to answer the technical enquiries that started to come in.

The Group still had neither money nor legal status. Its first donation came from Schumacher, who presented the Group with the fee he received for the *Observer* article; and part of that was used early in 1966 to set up the Group as a non-profit company and a registered charity, under the name of the Intermediate Technology Development Group Ltd.

How the new Group proceeded to pick up and act upon other indications of what people in developing countries were looking for has been described by Schumacher:

> It happened that a British Trade Mission was going to Nigeria, and we thought, my goodness, what are they going to flog in Nigeria? Certainly not the stuff the Nigerians really need. What if we made, very rapidly, a catalogue of hand and animal-propelled agricultural equipment obtainable in Britain? We had only a few weeks to do it, and under this pressure we found out how to touch the network; the network already exists. There is an association of agricultural engineers, one of producers of agricultural equipment. We got in touch with all of them and produced a little catalogue, stencilled, of animal and man-operated agricultural equipment obtainable from little hole-and-corner firms in Britain.
>
> When the mission went to Nigeria, they were most astonished. This thing was torn out of their hands by people who said: For the first time you bring us something of interest, instead of the usual glossy catalogues; and that gave us a clue as to what needed doing. We said, of course, this is a rough job, let's do a bigger job, not limited to agricultural equipment. And in 1967 we produced a big catalogue, *Tools for Progress: A Guide to Small-scale Equipment for Rural Development*. In 1967, after twenty years of aid, so called, which had involved many billions of dollars, this was the first time that somebody made a catalogue of suitable equipment for developing countries, and it caused a sensation. Although we were operating out of one small room, and had no sales apparatus, this catalogue rolled around the globe.[11]

We sold 7,000 copies and could have sold more. All the money we borrowed to print it was repaid.

Thus encouraged, we started to go into the subject at greater depth, and from then on our work programme began to take shape. We also began to formulate more clearly the convictions and insights that had brought the Group together in the first place, and which have informed our activities ever since. They are,

first, that the source and centre of world poverty lies primarily in the rural areas of poor countries, which are largely bypassed by conventional aid and development programmes;

second, that the rural areas will continue to be bypassed and unemployment will continue to grow, unless self-help technologies are made available to the poor countries with assistance in their use; and

third, that the donor countries and agencies do not at present possess the necessary organized knowledge of adapted, appropriate technologies and communications to be able to assist effectively in rural development on the scale required.

We emphasized rural development because we felt that unless the disease of poverty was tackled at its source, in the rural areas, outside the big cities, it would continue to manifest itself in three ways – mass migration into cities, mass unemployment, and the persistent threat (or actuality) of mass starvation.

The Group, not being an academic institution and looking upon itself as an action group, required a name, and we chose to incorporate the words 'intermediate technology' into our title. No name can do more than give a preliminary indication or hint as to the organization's main interest. The words 'intermediate technology' hint at two things:

that in matters of development there is a problem of technology, of choosing the right 'level of technology': in other words, that there is a *choice* of technology; that it cannot be assumed that what is best in conditions of affluence is necessarily best in conditions of poverty;
and
that the technologies most likely to be appropriate for development in conditions of great poverty would be in some sense

'intermediate' between (to speak symbolically) the hoe and the tractor, or the panga and the combine harvester. Thus in terms of capital cost per workplace, the intermediate technology lies between the almost nil cost of a workplace using traditional tools, and the £5,000 to £10,000-per-workplace of a typical rich country.

Know-how at this 'intermediate' level – and also the relevant equipment – obviously existed in many places, but no one could say what *gaps* there were, and there was no point anywhere in the world where this know-how could be obtained as and when the people most in need of it required it. Intermediate Technology *Development* therefore means the work of bringing this kind of knowledge to light, to systematize and, where necessary, complete it, and to organize a world-wide system of 'knowledge centres' where it can be readily found.

Nearly fifteen years of work in this field have supplied the Group with plentiful evidence that the 'knowledge gap' which it set out to fill is indeed very wide. The labour-saving, capital-intensive, highly sophisticated technologies, suitable for large-scale production in 'rich' markets, which are commonly used in the rich countries, are very well documented and easily accessible; but technologies applicable on a small scale by (or in) communities with plenty of labour and little capital, lacking technical and organizational sophistication, are, on the whole, poorly documented, difficult to get hold of, and in many cases even non-existent.

It will be apparent from what has been said that the Group, in insisting on *technological choice*, makes a sharp distinction between science on the one hand and technology on the other, or, to put it differently, between knowledge and its application. The knowledge of scientific principles, of 'laws of nature', of materials, and of methods is, in a sense, absolute, and to talk of 'intermediate knowledge' or 'intermediate science' would obviously be silly, implying a wilful rejection or disregard of human achievement. As far as knowledge and scientific achievement are concerned, the best is the best, irrespective of economic conditions. The *application* of the best knowledge, however, can take many different forms and can lead to many different types of technology and modes of operation. It is here that the need for, and the possibility of, intelligent *choice* enters: different economic conditions demand different applications. It cannot be doubted that the technological pacesetters today are the rich countries; probably something like 90 per cent of all research and development work (R. and D.) is done by them, in response to their needs and interests as they see them. Not surprisingly, therefore, the use *they* make of the achievements of science appears to many people as the only rational use. But this is an error: it may be the only rational use to meet their problem, yet a quite irrational use to meet the problems of poor people in poor countries. While there is something absolute in science, so that it can be rightly said that anything less than the most advanced is backward, there is (in this sense) nothing absolute in technology, which, to be fruitful, must fit the actual economic and social conditions within which it is intended to operate, and it cannot be said that the most 'advanced' is in all conditions necessarily the best.

The knowledge gap which the Group attempts to fill does not exist primarily at the level of science; it exists mainly at the level of appropriate technology, i.e., the application of knowledge, and technological availability. Here there is a crippling shortage of appropriate research and development. To overcome this shortage calls for a world-wide effort, which should be *systematic* but would be unlikely to be very expensive. The needs of poor people are quite limited and generally fairly simple; development tends to become self-generating when people have discovered, and have learned to utilize, realistic possibilities of self-help. One might say that, in the spirit of self-reliance, they will then pull themselves up by their bootstraps: but they cannot do so if they do not even have bootstraps. A systematic R. and D. (and communications) effort is required to provide them with bootstraps.

For this effort, only the best knowledge is good enough. In fact, experience shows that it often needs a higher level of creativity to advance the frontiers of knowledge conducive to intermediate technology, than is required in conventional R. and D. work. It is necessary to go back to first principles and to recognize constraints that exist in conditions of poverty and are absent in conditions of affluence. The road to the cheap and simple solution often demands the deployment of the most sophisticated thought processes and calculations, and can be successfully travelled only with the help of the latest and best research equipment.

To give more concreteness to the work programme indicated by the words 'intermediate technology', we emphasize four criteria: smallness; simplicity; capital cheapness; and non-violence. (Other people may prefer to choose a different set of criteria.) It is by no means certain that all four criteria can be satisfied in every case; *but any one of them, or combination of them, is of value for our purposes.*

The established trend of technological development is towards ever larger-scale, towards giantism. This is considered to be justified by the 'economies of scale'. But large-scale production units can be economical only when certain conditions are satisfied; a high market 'density', for instance, or a highly efficient low-cost transport system. When these conditions (and others, such as skill in large-scale organization, management, buying, and selling) are not satisfied, the so-called economies of scale become illusory. In fact, large-scale then tends to act as a principle of exclusion: only people already rich and powerful can embark on new productive enterprises; the small man is excluded, reduced to the position of a job seeker – and when there are not enough jobs provided by the rich and powerful, he has no possibility of becoming productive.

The importance of the criterion of smallness hardly needs to be argued, and experience shows that whenever efficient, small-scale equipment is made available the demand for it comes not merely from the Third World, but even more insistently from the affluent societies as well. Smallness is a *conditio sine qua non* for rural development, but it is also highly relevant from many other points of view – ecological, resource-wise, and social.

Much the same applies to simplicity and capital cheapness. It does not take great scientific or technological creativity to take a further step in the direction of complexity and capital-intensity. This means nothing more than following the established trend. But when the demand is made to search for smallness, simplicity and capital-saving, the normal first answer is that 'it cannot be done'. Experience shows that – not everywhere, but over wide ranges of application – this answer is simply wrong. It *can* be done, but it requires a more original R. and D. effort than is normally forthcoming.

Non-violence in this context refers to modes of production which respect ecological principles and strive to work with nature instead of attempting to force their way through natural systems in the conviction that unintended damage and unforeseen side-effects can always be undone by the further application of violence. All too often, one problem is 'solved' by creating several new ones.

Having given itself a name, and taken a two-roomed office in London's Covent Garden, the Group set itself the task of starting to fill the 'knowledge gap' by developing a series of programmes on technologies which are basic to rural life: agricultural tools and equipment, health and water supply, building materials and methods, energy, transport, small manufacturing and the like. In each case our object was to identify what was needed by way of improved equipment or methods, to discover ways of meeting these needs, to make the information widely available, and to demonstrate the feasibility of appropriate technologies by field tests and projects under actual operating conditions.

With no money but a growing number of friends and supporters, we started by drawing together, round each subject, a team of experts to advise us and help us to prepare specific 'fundable' work programmes. These voluntary panels rapidly became an integral part of the Group's structure, and by the end of the 1970s there were more than three hundred professional people – scientists, engineers, doctors, economists, men and women from industry, government, academia and the professions – helping us in this way. The most practical results, we found, generally came from panels consisting of people with different perspectives on the same subject: administrators, business people, academics and members of other voluntary organizations – what Schumacher christened the 'ABCD combination', in his last talk (see p. 12).

The first panel to get going, in 1968, was on building. It started not, as might be expected, with hardware development, but with 'software': with a training programme to upgrade the management skills of small building contractors in Africa.

The next subject to be tackled was small-scale water supply, and the water panel was soon into a series of field projects on rainwater collection and storage.

By the end of 1969 we had added similar panels on agriculture, health, and co-operatives. In that year too, the Group's first subsidiary company, Intermediate Technology Consultants Ltd, was started in response to the growing number of requests from international agencies, governments and other aid agencies for help in identifying needs and defining appropriate technologies. By charging fees for the services, the company was able to pay its own way and also contribute to the costs of the Group's charitable activities, the income for which had up until then come wholly as grants from foundations, voluntary agencies, companies and individuals.

By the early 1970s the Group had fifteen technical staff, of whom seven were in Zambia, Nigeria, Jamaica and Tanzania, running projects in agriculture, rural workshops, small-scale water supply and food technology. We had also started to publish the results of our first investigations and field projects, and were setting up another subsidiary company, Intermediate Technology Publications Ltd, to handle publications, and to produce a new quarterly journal, *Appropriate Technology*.

Today, with a permanent staff of more than fifty, the Group is working on a range of intermediate technologies spanning agriculture and water supply, building materials and methods, energy, transport and small industries. Supporting these programmes, and exploring a dozen other subjects ranging from chemical engineering to co-operatives, are some seventeen technical advisory panels. Between consultancies and field projects the Group is now typically working at any one time in at least twenty developing countries. A long and growing publications list documents the practical experience and knowledge of self-help technologies amassed by the Group and many others outside.

2

THE WORK OF THE GROUP TODAY

As an introduction to the main features of the Group's work, it is helpful to look at the way in which the work programme evolved – there was never any long-term plan.

The Group's task is essentially one of action research and development. This means that relevant technologies must not only be discovered or devised: they must also pass the tests of

filling an identifiable need;

being relatively cheap and simple to make and maintain; and

functioning satisfactorily under operating conditions.

These requirements have led the Group, from the outset, to associate as many qualified people as possible with its work. One aspect of this is the panel structure, which represents a broad spectrum of expertise – a combination of both theoretical and practical knowledge – that is required for this kind of research and development to be realistic. Another is that as far as possible, home-based technical staff operate from centres where the relevant knowledge and facilities exist. Thus the teams working on power, and on agricultural and water development, operate from a centre attached to the Engineering Department at the University of Reading. The work on building materials and methods is done at a commercially operated research and development unit associated with the Group at Cradely Heath, Birmingham. The team of

engineers working on small industry – Intermediate Technology
Industrial Services – is based in Rugby, in the industrial Midlands;
and the research officer on transport was at the Department of
Engineering Science at the University of Oxford until very
recently, when we created our own Intermediate Technology
Transport Company.

Building

It will be recalled that the Group's first panel was on building, and
that their first project was designed to find ways of upgrading the
efficiency of small local building contractors in Africa. No amount
of technological development, the panel argued, would be of much
use unless these small entrepreneurs could first of all develop the
management skills needed for them to take over from expatriate
contractors. But small builders were outside the reach of conventional training programmes, and this constituted a major gap in the
technological capacity of the construction industry – which, in
most developing countries, is second only to agriculture as an
employer of labour, forms an important bridge between agriculture and industry, and generally accounts for more than half of
the country's gross fixed capital formation. All over Africa there
was the same problem – growing construction programmes without the indigenous skills needed to execute them.

The Building for Development team, comprising an architect,
an engineer and an economist, carried out their field work in
Nigeria and Kenya. They experimented with different ways of
presenting management training material to the contractors:
through lectures, seminars, films, exhibitions, dramatizations.
They showed that the contractors, mostly men without any formal
education, *can* be taught these skills. The team went on to introduce the field-tested training methods into courses for contractors
run by the National Construction Corporation of Kenya, and
produced a series of teaching kits which the Group published in
the early 1970s.

The engineer member of this team, Derek Miles, remained with
the Group as a consultant. In 1976 he was helping the International Labour Office to run an African regional course on

construction management, held in Nairobi and attended by eighteen African countries. Two years later, and eight years after the building team had shown the way, the I.L.O. launched a two-and-a-half year programme for construction management training, in which twelve African countries are participating.

A parallel set of training courses is now being developed by the I.L.O. with the Group's involvement for Ministry of Works staff in African countries, on project management. The Group has also been advising the governments of Ghana, Gambia and Tanzania on national training programmes.

The building panel next turned its attention to construction materials and techniques. The small-scale, localized production of building materials has proved to be one of the most rewarding lines of investigation pursued by the Group. It branched off in several directions. One led to the investigation of mini-plants to produce standard Portland cement, a major programme which is being carried out by the Appropriate Technology Development Association in India. Their aim is to develop small production units which are competitive with the conventional large-scale – 1,500 ton a day – plants (see pp. 210–11, where the work of the ATDA, India, is described).

To complement this approach of scaling down the large-scale technology of cement, the building team also started to explore ways of upgrading traditional cementitious materials. Lime and lime-pozzolana mortars have a respectable history, which goes back for thousands of years. (Pozzolanas are minerals, usually of volcanic origin, which react with lime to set like a cement.) Here the object is to produce a standardized though comparatively low-grade cementing material which could partially or completely replace Portland cement over a wide range of uses. In many parts of the world it could, for example, largely replace cement for buildings up to two storeys high; lime-pozzolana mortars also have a much longer life than conventional cement.

A member of the Group's building team carried out initial experiments with the small-scale production of lime-pozzolana mortars in South India. As a result of this work the Group is now monitoring a pilot lime-pozzolana project in Arusha in Tanzania. This produces up to three tons of lime-pozzolana a day, for making building blocks, mortars and plasters.

Units of this kind can be locally built using local materials.

Compared with Portland cement plants they use only about half as much energy, and about one quarter of the capital, per ton of output. The scaling down of brick production has proved an outstanding success. A typical modern brick factory in an industrialized country produces about one million bricks a week. For a developing country this technology has several serious drawbacks. It is very expensive, costing upwards of a million pounds in foreign exchange to install, and running costs are high because spare parts, and energy, generally have to be imported. The technology is usually geared to a specific type of clay; and it takes at least two years to commission a new plant.

Highly efficient small-scale brick and tile units have been developed by the Group's building panel, under its chairman, John Parry, a building materials consultant. These small units produce not one million but 10,000 bricks a week, using hand-operated methods or very simple machinery. Capital costs per workplace are about £400 as against about £40,000 in a large modern brickworks. In small units, very large savings in fuel costs are possible by air-drying the bricks before firing, and local production virtually eliminates transport costs. The building team has set up plants of this kind in Ghana, Gambia, Egypt, South Sudan and Tanzania.

A detailed study of the comparative costs of different brick-making plants in developing countries was recently made by the David Livingstone Institute, Strathclyde University. This shows conclusively that intermediate technology based plants produce bricks costing half as much, or less, as bricks made in capital-intensive, mechanized plants. The study also revealed that this cost advantage held good even at a substantial scale capable of serving urban markets.[1]

One result of these practical demonstrations of intermediate technology in action was that the team began to be overwhelmed with requests for its services. Another was that new opportunities for developing low-cost materials suggested themselves. So they decided to set up a building materials workshop in which to conduct experiments and design and make prototypes.

They established themselves on the site of an old brickworks in Cradely Heath, near Birmingham. One of their first projects was to design and construct a transportable brickmaking kit. When complete this pilot unit proved to be capable of making 5,000 bricks a week from a wide range of clays and shales; it was easily transportable – it fits comfortably into a Land Rover – and the total cost of the kit was under £1,000. The kit includes a new design of mould that enables a trainee to make perfect bricks after only a few minutes' practice, and a mini-fork-truck that can carry up to fifty bricks at a time. The unit is supplied with an instruction manual in cartoon story form, and the kit is designed so that it can be copied by local craftsmen and be widely reproduced.

The mini brick plants are also versatile in their use of fuel. The workshop has devised ways of firing bricks and tiles using coal, firewood or waste engine oil.

Perhaps the most remarkable innovation of this building team has been in the application of fibre-reinforced cement to low-cost building components. Fibre-reinforced cement has, of course, long been used in the building industry. But the material most commonly used, asbestos cement, is manufactured in highly capital-intensive factories requiring special safeguards because of health risks from working with the material. Using a variety of animal and vegetable fibres, the building team have devised ways of making many fibre-reinforced cement components in a small workshop, with equipment easily available in any developing country.

The team have developed a very simple way of making corrugated roofing sheet out of fibre-reinforced cement. By this method a standard eight by three foot corrugated sheet can be made, with material costs as low as £2 a sheet. In many developing countries the equivalent asbestos-cement product, which has to be imported, can cost three times as much. Small pilot manufacturing units have been set up in Botswana, Gambia and Sri Lanka.

Trials carried out in Gambia showed that the cost of a locally made roofing sheet, including materials and labour, was less than half of that of the imported product. The same system can be used for other roofing components, and a complete 'roofing package' can now be used in a small workshop.

It can also be used for making fibre-reinforced panels for internal partitions, or for coating load-bearing walls made of stabilized soil or unfired clay bricks. Protective panels of this kind are under test in Nigeria and Gambia. Their potential for cutting down building costs has also attracted the attention of architects in

Britain, and similar panels are under test in Lancashire and in Devon.

Internal panels reinforced with bamboo and pine needles have proved so resilient that the workshop team are now experimenting with this method for making doors and furniture. Fibre-reinforced water catchment and storage tanks are another possibility now being investigated. Arrangements have been made with Hatfield Polytechnic for the testing of a variety of vegetable fibres, available in developing countries, to discover which are best suited for making fibre-reinforced cement compounds of different kinds.

An interesting feature of the building materials workshop is that its services are now being used by British manufacturers of building materials, who are becoming increasingly interested in simpler production methods as the costs of energy and capital equipment escalate. Contracts from British firms have included the pilot production of paving slabs using material formerly discarded as waste, and improved methods of making handmade bricks. The result is that the building materials workshop, which the Group initially financed, is now self-supporting.

Water

In the ten years that have elapsed since the Group's panel on small-scale water technology was formed, there has been a considerable shift in opinion about water supply. Thanks to the work of bodies such as the International Institute for Environment and Development, and many other voluntary agencies, there is now a much wider understanding of the importance of small-scale and self-help methods of getting water, both pure water for domestic use and irrigation water for smallholdings. This is in contrast with the prevailing attitudes of a decade ago, when the great majority of water engineers were preoccupied with building vast dams and reservoirs, and with high-cost Western systems of water supply and waste disposal. The late Michael Ionides, a brilliant engineer who foresaw the need for a small-scale water technology nearly a quarter of a century ago, used to say that conventional civil engineers regarded a river as a sort of challenge: how many dams could they build on it, how great a command area could be created, and so on. But for the vastly greater areas of the world's surface that are outside big river systems they had virtually nothing to offer.

In fact there is almost unlimited scope for intermediate technology in small-scale water development. Three-quarters of the rural people in developing countries have no access to safe water, and simple irrigation schemes are the only kind that are within the reach of tens of millions of peasant farmers and smallholders.

There are many parts of the world where it rains only for a short period each year, and where most of the rainwater runs to waste or is lost by evaporation. Every 100 millimetres (4 inches) of rain on 100 square metres (120 square yards) of land surface represents 10,000 litres (2,200 gallons) of water. The problem is how to collect and store this water. One way is by building rainfall collection tanks. The Group's work on water supplies started with a rainfall collection tank project in Botswana in 1967 and 1968, followed by similar projects in Swaziland, Jamaica, Brazil and Ethiopia. The collection tanks developed by the Group's water engineers are a good example of combining traditional techniques with modern materials: they use polythene sheeting, mud or clay, with sand and cement to line the tanks, and simple filtration methods can also be employed. Efficient tanks ranging in size from 10,000 gallons to three-quarters of a million gallons have been tested.

In order to provide itself with a research and development capacity, and to cope with a growing number of enquiries, the water panel set up a base unit with a full-time project officer in 1972, at the National College of Agricultural Engineering at Silsoe in Bedfordshire.

Following their work on water collection tanks, the water team has been engaged on several pioneering assignments overseas. Peter Stern, the Group's senior water consultant, spent two and a half years working with the Ethiopian government's water authorities between 1974 and 1977. During that time he identified and helped to implement some twenty small-scale water development projects in the drought-stricken areas of the Tigre and Wollo provinces. Each project comprised a variety of village level techniques including small collection dams, rainfall collection works, hand-dug wells, spring protection and minor irrigation development. Until then most the government's water development expenditure had been devoted to drilling expensive deep boreholes with imported equipment.

In the mid-1970s two other members of the Group's water team were working in Ethiopia. One was running a pilot project on self-help hand-dug wells in the Gurage region. The other was attached to the Surface Water Development Unit, which was set up by Oxfam to provide logistic support to the growing number of small-scale water developments throughout the country. This service centre designs and produces the hardware needed by small projects, arranges the purchase and supply of tools, and runs a mobile field operations and training team to give on the spot technical assistance. Within six months of its formation this unit was being run without any further expatriate assistance.

Much of the Group's experience and development of low-cost water technologies has been published in the form of handbooks on pumps, hand-dug wells, the hydraulic ram, water treatment and sanitation, and low-cost irrigation.

Farming

According to one of the reports presented to the United Nations Conference on Science and Technology for Development held in Vienna in the autumn of 1979,

> The developing world is overwhelmingly a world of smallness. Four-fifths of the farms are of five hectares (twelve acres) or less. Nearly half are just a single hectare . . . We now know from the history of a few countries, and also from experience with innumerable community programs carried on all over the world, that tools and machines can be designed for micro-enterprises that are more productive than traditional technologies, and that are low-cost and job-creating rather than labor displacing.[2]

This was certainly not common knowledge fifteen or even ten years ago, when the prevailing notion of agricultural scientists and engineers seemed to be that peasant farmers in the developing world had better either adopt Western methods and machines or go out of business.

Today it is quite obvious that the capital and energy-intensive farming systems of the West are singularly ill-suited to meet the needs of developing countries (and it can be doubted whether they are in any case sustainable). If the developing countries are to feed themselves, and it is a poor outlook for those that cannot, then ways must be found of enabling millions of peasant farmers to become more productive.

A great deal of effort, nationally and internationally, is now going into agricultural research. But it is still far too often the case that there is a huge gulf between what can be done within the research stations and what is actually being done by the farmers outside – unless they are very rich. This is not because the research results are wrong, but because they are simply not capable of being applied to the life and work of the smallholder – they are inappropriate for technological, economic or social reasons. In order to discover what improvements the small farmer both needs and can use effectively, much needs to be known about the farmer and his work throughout the annual cycle of farming (and community) operations.

One part of the agricultural panel's programme, which it started in 1969, was designed to identify the needs for improved small-farm equipment, and to get these tools made locally. There are periods in the farming calendar of all peasant communities when every available pair of hands is fully occupied: for example, in planting, weeding or harvesting. Between these peaks there are long slack periods, when there is a great deal of unemployment and under-employment. The selective mechanization of these 'peak' operations, the short periods when there is a labour shortage, can therefore raise productivity without creating unemployment. If, moreover, the tools and equipment needed to increase output can be made locally, during the slack periods, then at least one step has been taken towards bringing industry into the rural areas.

During a three-year project at Magoye, Zambia, the agricultural team identified these peaks, which are the constraints on greater production, by means of a detailed survey of how the community used its labour throughout the year. They found that the main problem in that case was transport, and to a lesser extent weed control and the harvesting of groundnuts. The project team's next step was to develop simple bullock carts – carts had not previously been used in that locality – various types of cultivators for weed control, and a simple machine for harvesting groundnuts. As there were very few craftsmen and blacksmiths in the area, they had to design a blacksmith's forge and show people how to build it, and use it, before local manufacture of the new farm equipment could be started. The work of designing tools and implements for local use, and teaching people how to make them, is continued in Zambia by the intermediate technology unit of Family Farms, a voluntary farm training and settlement project.

In the course of the Zambia project the team developed a survey technique which can be easily adapted for any other rural community. By simplifying the method they were able to employ local school-leavers to carry out the survey. There is, in fact, no substitute for this kind of farm-level survey if the real needs of small farming communities are to be discovered: you have to find out what the people are trying to do, and how they do it, before you can help them to do it better.

The results of the work in Zambia, including details of the survey technique, were published by the Group in *An Example of Farm Land Survey Technique Using Local Resources*.[3]

A parallel project was simultaneously run in Nigeria, in an area where weed control had already been identified as the chief bottleneck. The report of this work – *Farm Equipment Development Project, Daudawa, N.C.S. Nigeria*[4] – gives an account of how the team designed, developed and introduced small-scale equipment in collaboration with local farmers and blacksmiths. This included machines for weed control, for groundnut harvesting and for stripping kenaf, a local fibre.

Another part of the agricultural team's programme is to identify and make more widely known what already exists by way of simple farm tools and machinery. One of their most successful ventures is *Tools for Agriculture*,[5] a guide to low-cost farm equipment in commercial production. This proved to be very popular not only in developing countries but also in the United States – so much so that an American publisher has reissued the guide for smallholders in the States, with the addition of a list of many U.S. manufacturers of small farm implements.[6]

Most of the equipment in commercial production is, however, still beyond the reach of the very small farmer. So the Group also brings together and publishes detailed specifications of simple agricultural implements that are not on sale, but which can be made and maintained by small workshops and local blacksmiths. Nearly fifty items of such do-it-yourself equipment have been published.[7] They include several that have been designed and proved under field conditions by the agricultural unit. Besides the blacksmith's forge mentioned above, there is an animal-drawn weeding machine costing about one-sixtieth of its imported tractor-drawn equivalent, and a hand-operated metal bending machine that costs £7, one-tenth of the cost of a more sophisticated and mechanized version.

Many more small workshops are needed, especially in African countries, if the local production of farm implements, equipment for processing farm outputs, and the like, is to become widespread. The agricultural team has drawn on its field experience to produce a manual on rural workshops. This describes in detail the tooling up of small workshops, ranging from the basic tools for a one- or two-man workshop without power, through to the more sophisticated unit requiring power for wood and metal working.

The recognition that it is the small farmer who holds the key to world food production has led to new work on small farm technology both in rich and in poor countries.[8] The work of the Group and others such as the International Rice Research Institute has pointed the way as far as low-cost tools and equipment for the small farmer is concerned. But no one has yet seriously tackled the real threat that looms over the smallholder in developing countries, namely that he is being persuaded to adopt the chemical farming practices of the West. The intensive use of herbicides, pesticides and inorganic fertilizers is an advanced – possibly terminal – form of violence,[9] the equivalent in agriculture of the use of nuclear power in industry. In this sense the technologies of a sustainable, that is, a biological, non-violent agriculture, for small farmers in poor countries remain virtually neglected by the official aid and development agencies, to their great discredit.

Transport

Another of the basic needs of rural communities is better transport. Here the general trend, until quite recently, has been for much more research and development effort to go into ways of building roads than into improving the vehicles that travel on them.

The Group's transport unit is involved in both these aspects of rural infrastructure. Soon after the unit was formed in 1974, one of its consultants was working with the World Bank's Rural Access Road Programme in Kenya, advising on the local availability and procurement of tools for local road construction. In 1978 the Kenyan government accepted the transport team's specifications for such tools and their local manufacture.

This line of work has continued with the preparation, for the I.L.O., of a manual on tools and equipment for labour-based road construction. The transport unit is also supervising a major study of the nature and use of tools, equipment and local materials for labour-intensive public works schemes. This is also being done for the I.L.O. in collaboration with building research bodies in India, the Philippines, Tanzania and Tunisia, and the object is to promote local production of implements and construction materials.

The transport team is also doing research on basic vehicles. Early in 1978 the Bangladesh government asked for the Group's advice on ways of improving existing forms of non-motorized transport. This led to the Bangladesh University of Science and Technology starting work on the development of an improved cycle rickshaw with more efficient transmission and braking systems and a lighter frame, and on upgrading bullock carts and country boats; this is being done in association with the Group.

The transport unit is based at the Department of Engineering Sciences at the University of Oxford and works in association with Professor Stuart Wilson, a leading authority on bicycle transport. One of Stuart Wilson's inventions is the Oxtrike (so named because Oxfam initially funded the project), a highly efficient cycle rickshaw specially designed to be made in developing countries. It combines standard bicycle parts with a frame made mostly of sheet metal. Sheet steel is widely used in developing countries, and it can be cut and folded to shape with hand tools. Other innovations are a special three-speed gear, and powerful foot-operated brakes. At the end of 1979 the Oxtrike was under-going field trials in Malawi, Kenya, Zanzibar, Tanzania and India, with a view to its local manufacture. Meanwhile the transport team was producing a series of information papers on the wide range of pedal-powered and other low-cost vehicles in use in South East Asia, as a preliminary to further development work and to make known in other parts of the world what is already available in Asia.

The Group is dealing with transport on a considerably larger scale at its boatyard in Juba on the Nile, in Southern Sudan. This project is run by the Group on behalf of the Sudan Council of Churches. The boatyard makes specially designed shallow-draught ferro-cement boats, each forty-five feet long and capable of carrying up to twenty tons of cargo, in a region where other forms of transport are virtually non-existent. Twelve boats had been completed and sold by the end of 1978, new orders were running ahead of production capacity, and the project had become self-financing. Now that the enterprise is a going concern the Group is concentrating on training local management and technical staff to take over the enterprise.

Energy

When Schumacher was advocating the development of renewable sources of energy during the 1950s and 1960s, he was widely regarded as a crank. Today his foresight is acknowledged, and rich and poor countries alike are becoming increasingly interested in ways of getting useful energy from the sun, the wind, water, and vegetable material – biomass – (though many of us believe that in practice the amount that is being done in this field is negligible in relation to the need).

The Group's energy programme was started in 1968. From the outset it has been closely associated with the Department of Engineering Sciences at the University of Reading, and the energy unit operates from workshops on the university campus.

The Group's work on windmills for low-lift irrigation (three to ten metres, or nine to thirty-three feet) started when the Group's project officer, Peter Fraenkel, was asked to help with a windmill irrigation project on the lower reaches of the River Omo in Ethiopia. The riverside communities needed year-round irrigation to avert food shortages and periodic famines, but they had no way of raising water from the river to their smallholdings. A mission station in the locality was trying out Cretan windmills (with six or eight arms and triangular canvas sails) which are cheap and work, but at low efficiency. Peter Fraenkel introduced several modifications that greatly increased the water-lifting capacity of these locally fabricated windmills, and published the results of this project in *Food from Windmills*.[10]

This experience also revealed a major gap in existing windmill technology. The large windmills that are still widely used in Australia and the U.S.A. are far too expensive for small farmers and they cannot be locally manufactured. But there was nothing smaller that could compete with the standard (also in most cases imported) three to four horsepower diesel pump. The energy team set out to fill this gap. This was not simply a matter of scaling down the existing large windmill, because the smaller it is, the less efficiently it works; and its technology had remained unchanged since the nineteenth century. The team at Reading introduced a completely new transmission system (which changes rotary motion to the vertical motion needed for pumping). They also designed a new all-purpose rotor hub, which can hold up to twenty-four windmill blades, and thus offer a range of speeds to suit different pumping requirements; and they completed the package by designing a variable speed pump specially tailored for use with windmills. The result is a windmill pumping system that compares favourably in terms of capital cost and performance – and of course very favourably in terms of longevity and running costs – with the standard diesel pump; with the additional advantage that the I.T. windmill has been specially designed for local workshop manufacture. In 1980 prototypes were being field tested in seven countries in Africa and the Indian sub-continent.

The Group's aim is always to provide a range of options. For places where windpower is not the right answer for pumping water, a possible solution may be the Humphrey pump, which was first developed at the turn of the century but had been superseded by the diesel pump. The remarkable feature of this pump is its simplicity – it consists chiefly of pipework, and it can be made in any small workshop that has a lathe, a drilling machine and a welding kit. There are no gears or bearings, and the 'piston' is a column of water. But it took Reading University's team of engineers several years to arrive at a workable version. This is because they were scaling down a giant pump with a throughput of nearly two million gallons an hour, to a village level size capable of lifting some 5,000 gallons an hour to heads of up to thirty feet. This involved the team in having to work out the theoretical principles of the whole system, something the original inventor had not done. The mini-pump is considerably more efficient than a diesel pump of the same power (about ¾ horsepower) and like its giant predecessor it runs on gas. This means that it could be run on methane gas, in which case it would be independent of any fossil fuel. Two prototypes were being tested on behalf of the Group in Egypt and Nepal in 1980.

What amounts to a transformation of small-scale hydro-technology is also emerging from the activities of the Group's energy team. One development of quite extraordinary promise arose out of the design of a vertical axis, four-bladed windmill by Peter Musgrove, an aeronautical engineer. The Reading team hit on the idea of putting a similar design of windmill under water in a free-flowing stream. They tested one in the Thames and found that they had discovered a small, mobile and very powerful water turbine. The tests showed that a three-foot diameter 'underwater windmill' of this kind could produce as much power as a sixty-foot diameter traditional water wheel! This water turbine required the highest skills to design it; but it would be inexpensive and easy to make, and can be adapted to water pumping or electricity generation as required. Tests are still in progress, but it is certain that we shall be hearing much more about the water turbine in the next year or so.

No less dramatic in its potential for small-scale hydro is a device invented by Rupert Armstrong-Evans, one of the Group's consultant associates. Previously, electricity generation from small-scale water flows was inhibited by the high cost of the equipment required, especially by the high cost of the governor, which controls the volume of water passing through the turbine. Armstrong-Evans has invented a relatively simple and very low-cost electronics device which controls the output of electricity rather than the intake of water, and thus eliminates the need for the expensive governor. The result is that costs of small hydro installations can be reduced by up to two-thirds. This electronic load control device is already operating in some thirty hydro sites in Britain, and is now undergoing field trials in Nepal, Pakistan, India, Sri Lanka and Fiji.

Another way in which the same inventor is widening the horizons for mini-hydro systems is by devising turbines that can operate on much lower heads of water than was formerly possible. By using a turbine that looks very like a ship's propeller – but is easier to make – he can generate power from heads of water as low as two or three feet.

These two inventions should soon make possible a major expansion in the use of mini-hydro installations, capable of producing enough power for, say, a village of a thousand people to pump their water supplies, run agricultural processing equipment, power a sawmill or workshop, and do some irrigation.

They are also, incidentally, excellent examples of technologies that are capital-saving. Mini-hydro plants that can be installed at a cost per kilowatt of between £100 and £300 compare favourably with modern thermal power stations in terms of capital costs, but of course the mini-hydro system gets its fuel free, and there are virtually no transmission costs.

While the industrialized countries have brought the whole world within sight of a permanent oil shortage, the poor countries have their own energy crisis in the form of a shortage of firewood. Wood is used for cooking and heating by the great bulk of the world's population, and wood-fuel is becoming increasingly scarce.

> Although rising oil prices have commanded newspaper headlines, rising firewood prices have fueled inflation in countless Third World countries. Price rises of two, three or fourfold during the seventies have not been uncommon. In Niamey, the capital of Niger, firewood purchases absorb a quarter of a manual worker's wages. In Kathmandu, Nepal, the price has tripled over the past few years; in Andean villages in Ecuador, many families are reduced to one hot meal a day. In some deforested Third World communities, what goes under the pot costs more than what goes inside it.[11]

Experiments in a number of countries, including Botswana and Guatemala, have shown that traditional methods of cooking and heating can be vastly improved by the introduction of better designs. Towards the end of 1978 a member of the Group's Reading team started to collect data on wood-stoves, with the aim of developing and testing different stove designs and combustion systems so that a package of information, advice and training can be made available to field workers overseas. This is being done in collaboration with interested organizations in a number of developing countries in Africa, Asia and elsewhere.

Health

In another example of the Group's approach and method of work, we describe an appropriate 'technology' which is less concerned with hardware and more with the proper use of human skills: health.

In developed countries a change in the lifestyle of the middle-aged would contribute very substantially to the saving of lives. In the U.S.A. half the deaths are caused by diseases of the heart and blood vessels and occur in people aged fifty to seventy. Diseases in developing countries are largely caused by poverty and they would virtually disappear if the people had a clean water supply, adequate housing, sanitation and education. Half the deaths are of children under five years. According to the World Bank, one-fifth of the population of the world, 800 million people, has no access to even minimal health care.

In poor countries health care cannot be divorced from general social and economic improvement. Almost everything that happens affects the health of the people. A pure water supply can drastically reduce the incidence of water-borne diseases, so more people are fit and able to work during the sowing season. A larger labour force during this critical period can greatly increase production and thus improve nutrition. The building and maintenance of the village health post creates employment and introduces villagers to sanitary latrines. A new fuel-efficient stove design may cut down the incidence of bronchitis and eye disease. An improved building technique may lessen insect infestation. New roads, besides helping in the marketing of cash crops, enable community workers to reach the villages.

In most developing countries health care systems have been copied from the West, where health care is in the hands of doctors and is hospital-based, centralized, dominated by very advanced technology and very expensive. Poor countries, with perhaps $1 or 50p per head of the population to spend on health, are looking for a way to spread health care to as many people as possible. Doctors trained in the Western tradition are concerned for the individual, not the community. There is insufficient emphasis in their training on preventive medicine, public health and health education. They are unwilling to work in rural areas; indeed, many are unwilling to work in their own countries at all, and they emigrate.[12]

It was against this sort of background that the Rural Health Panel was drawn together in 1968 by the late Professor Kenneth Hill to look for more appropriate, equitable and accessible systems of health care. The Panel's approach was in line with the ideas of intermediate technology: the work to be done must be simplified and then broken down into small parts, so that each can be taught to someone fairly quickly.

> Within every population, whatever its educational level, suitable recruits can be found and trained as medical auxiliaries to provide at least a better standard of health care than is now available to those most in need of it.[13]

However, the Panel's view of a medical team headed by the doctor with a wide range of trained helpers was not well received at that time, either in the profession or by the W.H.O. A senior W.H.O. official, writing in the *Journal of Medical Education* in 1969, saw health assistants only as a very temporary solution and reacted strongly against 'hasty half-trained medics' and the 'dilution of medical education'.

Health auxiliaries have a long and respectable history. Peter the Great introduced *feldshers* into Russia in 1700. The Sudan and Fiji have been training people to help in the villages for nearly one hundred years. Former British and French colonies in Africa had auxiliary medical workers over fifty years ago, and India had its Licensed Medical Practitioners or Surgeons' Assistants. Ethiopia and Papua New Guinea have trained auxiliaries for over forty years. The 'barefoot doctors' of China have been going strong since 1964.

The training, which is given at very many different levels across the world, varies from three weeks for some Kenyan village women workers, who learn about hygiene, child care and nutrition, to three years for medical assistants in Tanzania and Papua New Guinea. The key to all training is relevance, and training is by doing.

Most schemes for using health auxiliaries operate at three levels. The health centre, which may serve some 150,000 people, acts as a referral centre and has a rural hospital. It is also the training centre for all the different levels of auxiliaries. The staff in charge may be doctors or they may be medical assistants with three to five years' medical training. There may be seven or eight sub-centres, often called rural dispensaries, and each will serve about 15,000 people. Some of the staff will have a three-year medical training, some will concentrate on maternal and child health and family planning, some will deal with environmental health and sanitation, and others will run the immunization programmes. Village health posts are run by village health workers who are villagers themselves, chosen and paid by the villagers and willing to live and earn at their level. They get three to six months' training in basic knowledge of the diseases prevalent in their area and in nutrition education, maternal health, care of the under-fives, family planning and environmental sanitation, and they are given frequent refresher courses. The village health workers' task in the front line of health care is to encourage the community to look after its own health, and this they do by sharing their knowledge and setting an example.

In June 1971 the Health Panel published *Health Manpower and the Medical Auxiliary*.[14] This was the first-ever annotated bibliography on various aspects of the work of medical auxiliaries, but it also served admirably as a manual or guide to the economic and medical advantages of using them. It filled a large gap in the literature available to medical administrators and health planners in developing countries. It also revealed the great dearth of teaching material on the subject. In 1975, Dr Katherine Elliott, the Panel Chairman, compiled *The Training of Auxiliaries in Health Care*.[15] This was an annotated bibliography of auxiliary teaching materials, which included information on text books, course descriptions and visual aids for use at different levels. It was the first attempt to bring together information on the training of auxiliaries and it was very well received. A revised and expanded edition, *Auxiliaries in Primary Health Care*,[16] edited by Dr Elliott, was published in 1979 and carried an enthusiastic foreword by the Director-General of W.H.O., Dr Halfdan Mahler.

Dr Elliott's work, from 1975, for the International Hospital Federation on health auxiliary practice in different parts of the world, led to the publication of *Health Auxiliaries and the Health Team*,[17] which discusses the training, role and use of health auxiliaries in developing countries and in the U.S.A. It also led to the idea for a special Health Group to take over from the Rural Health Panel. Meanwhile, the W.H.O.'s Appropriate Technology for Health programme had begun and they invited the London group to become the first W.H.O. Collaborating Centre and offered some money in support. This was the beginning of the Appropriate Health Resources and Technology Action Group (AHRTAG), the successor to the Panel, which now functions as a separate non-profit making organization, working closely with I.T.D.G. It offers specialized services to health planners, educators and field workers to promote the appropriate and effective delivery of primary health care, including the development of appropriate health-related technologies, in all parts of the world. Dr Elliott is Honorary Director.

In collaboration with the W.H.O. Expanded Programme on Immunization (E.P.I.), AHRTAG established in 1978 an information centre for the 'cold chain' – the safe cold storage and transport of vaccines in tropical conditions. It is also involved in E.P.I.'s training programmes for 'cold chain' workers. The effective use of health auxiliaries continues to be held back by the lack of teaching materials specially designed for use at village level. Professor David Morley, a former chairman of the Rural Health Panel, founded Teaching Aids at Low Cost (TALC) in 1969 to help meet this need. TALC produces books and pamphlets, charts and colour slides covering the identification and management of a range of common health problems and very simple pieces of equipment or instructions for making them. AHRTAG plans to extend its work into this field and to produce a range of teaching materials for use by village health workers.

The effect which trained auxiliary health workers can have on the health of village people is well illustrated by two projects which were started in Bangladesh after independence. An evaluation team visiting one of those, the Gonoshasthaya Kendra community health project, in 1977, found no skin or diarrhoeal diseases, no smallpox and no maternal deaths. Birth and death rates had both dropped.[18] Results from the other, the Companiganj Rural Health Project, show improved nutritional standards, that malaria and smallpox is being controlled and that family planning users have increased by 20 per cent. Perhaps the most interesting finding here is the special success and effectiveness of some illiterate or poorly educated women in persuading fellow villagers to bring children to the clinic and to use family planning.

Many projects insist that villagers take responsibility for their own health care. At Lardin Gabas, Nigeria, *all* villagers are required to contribute time, energy and money to build and maintain their health centre, and drugs and equipment are not forthcoming unless there is wholehearted co-operation.

People are coming to realize, too, that some traditional medicine is effective. In Asia and elsewhere the use of traditional medicine has long been an established science. The W.H.O. has a working group on traditional medicine and has held seminars in Asia and Africa to encourage local healers and midwives to upgrade their knowledge and to collaborate with the national primary health care services. Local healers can be particularly valuable in dealing with mental health disorders which, as in the West, account for one-fifth of health problems.

In the examples given above, the Group has succeeded in raising sufficient funds to employ full-time research and development staff or, as in the case of the Building Materials Unit and AHRTAG, the units themselves have become self-supporting. These units are, so to speak, the front line of the Group's activities. Most of them are supported and advised by voluntary panels.

Even when it has not proved possible to appoint full-time staff or get R. and D. units operating, panels support the Group's work by helping to answer technical enquiries, preparing 'state of the art' surveys on their subjects, and producing publications. The Chemistry and Chemical Engineering Panel, for example, helped the Technology Consultancy Centre in Ghana to start its small-scale soap project, and continues to advise it on matters such as the local production of alkali. The Forest and Forest Industries Panel has been working on a proposal to develop and test a prototype mobile sawmill, possibly pedal-driven, as an improvement on traditional pit-sawing. Two of its members have designed and built prototypes of a solar kiln that can dry 9 cubic yards of timber. This is made in the form of an easily transportable kit. The Co-operative Panel has produced a series of programmed learning textbooks specially written for the use of members and staff of co-operatives in developing countries. These have been published and widely distributed by I.T. Publications.[19]

Women

It is now accepted that increased development in the rural areas of Africa and other parts of the Third World cannot possibly be accomplished while the technological needs of the women, who form 60 to 70 per cent of the agricultural labour force, continue to be ignored.

The woman, who is the farmer and overall provider for the family and not merely the farmer's wife, must be given at least equal access to improved tools, labour saving devices, training and credit. Rural African women have themselves specified that they need most help with fetching and carrying water and firewood, and crop processing.

In 1975 a senior member of the I.T.D.G.'s staff, Marilyn Carr, working with the U.N. Economic Commission for Africa, took on the task of helping governments to identify and promote the use of appropriate technologies for African women, with a view to making the tasks of rural women less burdensome and more productive. This was an important step in the movement towards a more effectively run society in rural Africa. The African Training and Research Centre for Women, under whose aegis Marilyn Carr worked, continues to produce very valuable research in this field.

The role of women as producers has been very largely neglected by aid and development programmes:

> Modern equipment which has been introduced has, almost invariably, been aimed at men and resulted in more rather than less work for the women. For example, partial mechanization helps the men to clear larger acreages of land with less effort, but the women are left to weed and harvest the enlarged area with traditional implements.[20]

The introduction of intermediate technologies specially designed for the work done by women is bound to have an immediate effect on family health and living standards. Thus, a cheap milling machine that enables a woman to grind maize into flour can save hours of tiring pounding. A pure water well near by could reduce sickness and save hours per day spent fetching water. A simple hand-operated weeding machine could save hours of drudgery in the fields. Time thus saved could be spent on child care, home improvement, or growing vegetables with some protein, to improve the family diet.

In many cases, simple improvements to existing technologies are better than anything which could be imported. With this in mind the E.C.A. Research Centre for Women funded surveys of indigenous technologies that could be improved to provide new jobs and enhance local and national self-sufficiency. The first surveys were carried out in Sierra Leone and Ethiopia: others are planned in Ghana, Morocco, Sierra Leone, Egypt and Tunisia. The surveys identify the role of women in operating village level technologies, and describe the operations, materials and equipment involved in activities such as oil extraction, dye preparation, fish smoking, various forms of food processing and preservation, and agricultural work.

In some cases, better ways of doing things already exist and the need is to find ways of making them available to village women. Thus simple methods of rainwater collection could often eliminate water-carrying, a task on which women in many areas have to spend as much as four hours on a single journey. But the problem is how to store the water once it has been collected:

> Conventional storage tanks, usually made out of metal, are beyond the means of the average family. Work is being done, however, on the development of cheaper storage containers. One of the more promising technologies is a thin-walled cement jar originally developed in Thailand and which is now used in almost every household in that country. The cost of the cement used in making a jar with a capacity of 3,000 litres is about 10 dollars. This compares very favourably with the conventional galvanized iron container of the same size which costs 100 dollars. The cement jar has now been introduced in Africa through the Village Technology Unit in Kenya.[21]

Better agricultural tools have been developed that could save women from much stooping and bending and hard physical effort.

These include hand-operated weeders and seeders, and low-cost threshing and winnowing machines.

But in other cases no complete answers have yet been found. Improved stoves (as we have seen earlier) can cut the use of firewood by half, but firewood still has to be found and carried. Methane gas or solar cookers may be a solution, but both still require a great deal of development. In some instances women could, once shown how, make the equipment themselves. A case in point is a simple solar drier, which can easily be made from mud, wood and polythene, and which could enable women to preserve surplus vegetables for home use or for sale throughout the year.

Among the pilot projects recently launched by the Research Centre for Women there is one on starting up consumer and production co-operatives for rural women in Ethiopia. This is part of a programme for training women how to make the most productive use of the time they save, after new wells have been constructed in their villages. A food preservation project in the Ivory Coast will identify improved ways of fish smoking and teach the new techniques to women. Another in Mauritania is introducing simple date pitters (adapted from American cherry pitters) to nomadic women. A project in Sierra Leone is aimed at installing hand-operated oil presses in selected villages in Sierra Leone. The presses were developed and made by the Engineering Department of the University of Sierra Leone. Sierra Leone is also being helped to start a village technology testing and demonstration unit. A variety of post-harvest and crop processing equipment is being introduced to pilot villages in the Gambia and Upper Volta.

The opportunities for new work on women's technologies that Marilyn Carr helped to open up have led UNICEF to appoint an officer in eastern Africa, with specific responsibility for women and the promotion of rural technology. UNICEF has also funded Marilyn Carr's successor at the E.C.A. Research Centre for Women. In terms of its potential for raising the standard of life in African rural communities through self-help technologies, this is likely to be the most productive venture which the Group has ever helped to get started.

Growing Support

Until the mid-1970s there was very little support from aid-giving governments or international agencies – the I.L.O. was an outstanding exception – for the idea of intermediate technology, or for the organizations that were showing that it could work. The latter included our Group and also Volunteers in Technical Assistance (VITA) in the U.S.A., the Brace Research Institute in Canada, and a growing number of voluntary agencies the world over. Then the scene began to change. The era of cheap oil had ended. Unemployment and poverty in the Third World continued to spread at alarming rates. It was becoming increasingly difficult for anyone to avoid the conclusion that wherever the salvation of poor countries and communities might lie, it was certainly not in the slavish copying of the capital- and energy-intensive technologies – and their accompanying lifestyles – of the West. This realization was driven home as more became known about China, where full employment had been achieved by a national policy that insisted on local self-reliance and the use of intermediate technologies.[22] Governments and international agencies began to give support to work on appropriate technology, and – especially after the publication of *Small is Beautiful* – appropriate technology groups and organizations started to spring up in rich and poor countries alike.

This change in the climate of opinion about the role of intermediate technology in development is, of course, far from complete. There are many people still arguing that intermediate technology is a second best, or worse still, an instrument of neo-colonialism, without pausing to consider the question: second best for whom, the people who are already rich, or the very poor who have nothing? Or to reflect on whether neo-colonialism is better represented by technologies of self-reliance, or by technological and economic dependence on rich countries. But the process of reappraisal has begun and is likely to lead to major changes in development objectives and strategies, especially among the poor countries.

For the practitioners of intermediate technology on both sides of the Atlantic the change that has already taken place has opened up new opportunities and new ways of working. In the first ten years of the Group's existence, we found ourselves both advocating a case, and having to prove it by setting up our own field projects. From the mid-1970s we were able to spend less time arguing the case and more effort in showing what can be done, in collaboration with others who are now working in the same direction.

Support from the British government, for example, arrived via the Group's work on small-scale manufacturing industry. We started this in 1969, when we formed a small industrial liaison unit and based it in the Midlands. Its aim was to draw on the expertise of small-scale industry in Britain and adapt it to the needs of small firms in developing countries.

One of this unit's overseas projects was the intermediate technology workshop in Zaria, in northern Nigeria, where young Nigerians were trained to manufacture a wide range of equipment, formerly imported, for use in Nigerian hospitals and clinics. Soon the engineer who started that workshop, the late Will Eaves, was running a much larger and highly successful production/training programme in northern Nigeria (see pp. 236–8).

Another arm of the Group's industrial unit, meanwhile, was engaged on the design and manufacture of a mini-machine for paper pulp packaging. This project arose out of a visit to Zambia by two of the Group's directors in 1969. Zambia – and as it later transpired, many other countries – was looking for a machine to make egg trays from waste paper, on a small scale, to suit their markets. They needed about one million egg trays a year. But the smallest machine then available produced a million a month. To fill this technological gap the Group worked with the Royal College of Art, the University of Reading, and a small commercial manufacturer, and finally came up with a mini-plant capable of making about 600,000 egg trays a year, and which could be scaled up as required. In 1978, after more than a dozen other countries had bought and installed these mini-plants, and they had been thoroughly proved both technically and commercially, the Group handed over the manufacture and marketing on licence to a commercial company (the one that had done the earlier development work on the machine). It is now exploring the possibility of manufacturing the mini-plant in a developing country.

The interest aroused in developing countries by this kind of work encouraged the British Ministry of Overseas Development to finance the Group on a larger scale. This support enabled us to set up a new unit, Intermediate Technology Industrial Services, in

1978. ITIS is a team of some twenty engineers whose task is to discover and develop, test, and help to introduce small-scale manufacturing technologies in developing countries. This has added a new dimension to the Group's ability to respond to technical enquiries – which were running at the rate of about 1,000 a year by the late 1970s – and to carry through to commercial operation new or adapted technologies that arise out of the Group's own researches, or out of the work of other organizations and individuals at home or abroad.

By the end of its first year the ITIS team was working on thirty projects, ranging from feasibility studies of solar pumps, a mini-glass plant, and small electric motors, to the field testing of roofing components and improved methods of wool spinning. They had also started collaborative projects with Indian firms on small-scale papermaking, with the Technology Consultancy Centre in Ghana on the setting up of a small foundry, with a group in Nepal working on mini-hydro, and with the Philippine Centre for Appropriate Training and Technology on the testing of low-cost salt evaporating plant.

Internationally, the growing recognition afforded to low-cost technology for rural development is reflected in the number of appropriate technology organizations that are starting up in the developing countries themselves. Obviously, if intermediate technology is to serve its purpose, it must become a normal, accepted part of a country's development, and not something that remains exceptional. It must become, that is, a part of a country's indigenous culture. This can only start to happen when there are organizations within each country that can identify the needs of their communities, and find ways of meeting these needs in collaboration with local people.

Within a year or two of its formation, the Group was assisting the start of appropriate technology (A.T.) units overseas, in India and then in Ghana. Since then we have been associated with the formation of several others, notably in Pakistan, Botswana and Lesotho.

The Group's efforts to build up a sustainable capacity for self-help within the developing countries were greatly reinforced in 1975, when the U.N. Economic Commission for Africa created two posts which they asked the Group to fill. One was for an African regional adviser on A.T. and small industries: the other

was for an expert on village technology to be attached to the
E.C.A.'s African Training and Research Centre for Women (see
p. 60).

The small industries adviser was David Wright, who later
became General Manager of ITIS, on its formation. During his time
with the E.C.A. he was closely involved with appropriate technology development in Botswana, Lesotho, Malawi and Kenya, all
countries in which the subject of appropriate technology has
become an important part of national policies. The government of
Botswana has since set up a technology centre with support from
the European Development Fund. Lesotho has started Lesotho
Technology and Consultancy Services as a subsidiary of their
national development corporation, with funding by the Commonwealth Fund for Technical Co-operation and Appropriate Technology International of the United States.

An International Network

The I.T.D.G. is, of course, only one of a growing number of
organizations which are now helping to furnish people in developing countries with the technologies of self-help.

To begin with Britain, work on appropriate technology is no
longer restricted to the voluntary, non-government sector. The
Commonwealth Fund for Technical Co-operation for instance, has
started to sponsor the exchange, between African countries, of
technologies appropriate for the small farmer. This began with a
meeting of countries from East, Central and Southern Africa at
Arusha, Tanzania, where machinery from around the region was
displayed and compared.

To strengthen this exchange of experience and know-how,
National Appropriate Technologies Committees have been set up
by the governments of Zambia, Kenya, Malawi and Tanzania, and
others are in prospect for Swaziland and Mauritius. Already fruitful exchanges are taking place.

A water pump, developed in Malawi and shown at Arusha, has
aroused keen interest in all of the countries and Tanzania is
particularly keen to follow Malawi's lead and use it for village
boreholes. Swaziland's small tractor, the Tinkabi, has now been

ordered by Zambia, and Malawi and Kenya both want to assemble it locally. A workshop on the Tinkabi tractor will take place shortly, visits to water-pump projects in Malawi, and village blacksmith programmes in Tanzania are being arranged, and a general African workshop on dryland farming is planned.[23]

The countries involved in these exchanges have since asked the Commonwealth Secretariat to prepare a guide to technology transfer in the region, including a catalogue of implements available, details of transport routes and other aspects of technological exchange. The results of this initiative by the Commonwealth Fund for Technical Assistance are very encouraging and it seems likely that this kind of collaboration will spread among other members of the Commonwealth.

Appropriate technology is also getting its foot into the door of government research establishments. One of the research and development units of the Ministry of Overseas Development is the Tropical Products Institute (T.P.I.), an organization of international repute. A growing proportion of the 2,000 enquiries they get each year is now reaching the T.P.I.'s industrial development department, which is working on a range of low-cost technologies for post-harvest processing in the tropics. Their chief emphasis is on food processing: getting food into a less perishable form, or into the right condition for storage, or into a marketable product. As we have seen, this is the kind of technology that is essentially appropriate to the needs of women in developing countries.

Examples of this kind of equipment that are already in use or being field tested include a wooden maize sheller that can be made in any African village, a pedal-operated mill that can grind hard grains to any degree of fineness, and a range of small-scale decorticators, hand-operated and easy to make locally, to remove the husks from cotton and sunflower seeds. They are also working on the small-scale extraction of oil from seeds such as rape and sunflower.

Elsewhere in Europe, support for appropriate technology is part of the official aid programmes in France, Germany, the Scandinavian countries and the Netherlands. In Germany the Federal Ministry for Economic Co-operation has set up a separate organization, the German Appropriate Technology Exchange, to handle all aspects of appropriate technology for developing countries. In France there is the officially sponsored Groupe de Recherches sur les Technologies Appropriées, a small information unit which is also responsible for the training of volunteers working in developing countries. The Netherlands is the home of TOOL, which started in 1974 as the central organization of a number of groups in Dutch universities and colleges which were answering technical enquiries from people working in developing countries. TOOL is the focal point of some 600 overseas requests a year, and is building up a series of supporting research and development projects. In 1978 the organization was working on a low-cost wooden fish-net weaving machine; an improved evaporator for small-scale sugar making, in collaboration with Amadu Bello University in Nigeria, and a solar powered ice making unit. Another of its teams is working with the Allahabad Polytechnic and the Organization of the Rural Poor in north India on windmill development.

On the other side of the Atlantic there are now several hundred groups working on appropriate technologies. Most of these are in the U.S.A., and the great majority – as we shall see later – are directing their efforts towards changing technologies and lifestyles in the United States itself. Among those that are orientated towards developing countries, two, Volunteers in Technical Assistance (VITA) in the U.S.A., and the Brace Research Institute in Canada, are pioneers in the field of low-cost, small-scale technology development. Another, much more recent, organization is Appropriate Technology International (A.T.I.), which represents an unequivocal acceptance by the U.S. government of the importance of intermediate technology for the developing world, and a major commitment to the task of building up self-help organizations within the developing countries.

VITA was formed in 1960. Its basis of operation has been an enquiry service which draws on the scientific and technical knowledge of no fewer than 4,500 technical experts throughout the U.S.A., to provide information and advice to individuals and groups in the developing countries. While at one time a high proportion of these enquiries came from expatriates working with voluntary organizations, the bulk are now requests from indigenous groups and individuals. VITA, like I.T.D.G., is now increasingly working with local organizations overseas, by attaching technically qualified volunteers to specific projects and programmes. They are collaborating, for instance, with the Société Africaine d'Etudes et de Développement (SAED) in Upper Volta to set up a technology development and documentation centre; they are also helping SAED to develop improved designs of tricycles, motor cycles, pumps, and methane digesters. In Papua New Guinea VITA is working with, among others, the South Pacific Appropriate Technology Foundation on small-scale water technologies. One of the outcomes is that a very efficient, low-cost hydraulic ram pump is now being made locally, and work continues on small-scale hydrosystems. They are also helping the Liklik Buk information centre, run by the Melanesian Council of Churches, with information dissemination.

VITA are providing the technical expertise for a project run jointly with the Centro Cooperativo Tecnico Industrial (C.C.T.I.) in Honduras to improve lime and salt production, and successful pilot projects for both have been set up. Other VITA volunteers are working with the International Institute for Tropical Agriculture in Nigeria, helping to design and make low-cost equipment such as manually operated seed planters, threshers and sprayers for use on small farms. VITA are also advising the World Bank on potential appropriate technology contributions to the bank's programmes.

With nearly twenty years of experience in responding to overseas enquiries behind them, VITA have built up a very efficient system for information storage and retrieval, and their publications unit produces guides, manuals, and leaflets on low-cost technologies that are distributed at the rate of some 20,000 a year.

Much smaller, but with no less of an international reputation, is the Brace Research Institute, a research and development unit of MacDonald College, McGill University, Canada. It was started in 1961 and has become one of the leading international research centres for solar energy and wind energy utilization, and especially on solar distillation and wind turbines. As might be imagined, their work is now attracting a great deal of attention within Canada, where they have made major contributions to the development of efficient, low-cost greenhouses, and solar-heated buildings. The main part of their work, however, continues to be directed to the needs of poor communities in developing countries.

The Brace team is currently advising the University of Dar es Salaam in Tanzania on the use of renewable energy technologies, including solar cookers and food warmers. They have recently provided an appropriate technology group in Colombia with plans for simple forms of desalination, needed in Guajira, one of the most arid regions in the country, and guidance on the future development of solar and wind energy for the region. They have also made design studies for renewable energy systems to meet the water and energy needs of rural communities in the semi-arid areas of northern Venezuela.

One of the Institute's most imaginative projects has been a feasibility study of ways of meeting the energy needs of villages in Senegal. They have published this as a detailed plan to provide the water, cooking and lighting needs of a small community of about 500 people, typical of the Sahelian region of Africa. The scheme involves the use of windmill pumps, solar cookers and biogas (methane) digesters, and is designed to use low-cost technologies in ways that integrate with village life.

The Institute's work on this scheme revealed that 80 per cent of the basic energy needs in such areas is for cooking fuel. There is an urgent need to minimize the use of wood fuel, and the Institute has made the development and testing of different models of low-cost solar cookers an important part of its work on renewable energy devices.

The growth of the appropriate technology network, linking centres such as I.T.D.G., VITA, Brace, TOOL and the others in the industrialized countries, and local and national organizations in the developing countries, has recently been reinforced by the creation of Appropriate Technology International. A.T.I. is a private, non-profit making corporation, based in Washington and funded – in 1978 to the tune of some $5 million – by the U.S. government's Aid Administration. A.T.I.'s mandate from the U.S. Congress is 'to promote the development and dissemination of technologies appropriate for developing countries.' They have been given a free hand to undertake this task, and they have started in a very promising way by putting all their efforts into providing financial assistance to developing countries to strengthen indigenous capacities to develop, adapt and employ appropriate technology.

Within a year of being formed the A.T.I. were supporting more than a dozen organizations in ten developing countries. Some are technology user groups which need better access to credit or capital, such as a community co-operative on the outskirts of Nairobi which comprises small construction and metal-working firms, a sewing shop, handicrafts, and a small farmers' co-operative; or the group of sixty co-operatives in Nicaragua, half of them women's market co-ops, that want to expand their membership and credit base. Others are organizations that need assistance in developing or introducing new technologies. A.T.I. are providing financial support for the Appropriate Technology Development Association (ATDA) in India to develop a mini-cement plant; an appropriate technology group in Guatemala – the Centro Meso Americano de Estudios Sobre Tecnologia Apropriada (CEMAT) – to disseminate the construction of an efficient wood-burning stove; the Arusha Appropriate Technology group to expand its extension and training work; the Lesotho government to set up a technology centre for small business development, and to introduce a micro-hydro programme.

In their second year, with a total of more than forty overseas projects in hand, A.T.I. were helping to build up more than a dozen A.T. information and resource centres in Latin America, Asia, Africa and the South Pacific. It would be difficult to think of a more useful activity at the present time than the building up of organizations in the developing countries that are specifically concerned with the development and dissemination of appropriate technologies. It is only when locally based groups and organizations of this kind can show practical results, that the governments of developing countries are likely to start building appropriate technology into their development programmes – and, what is equally important, insisting that the aid-giving countries offer them a range of technology from which *they*, the developing countries, can choose what suits them best.

During the past few years, many A.T. organizations have been set up in the developing countries.[24] They are mostly small groups lacking in resources, but some have already built up really impressive work programmes. An account of the work being done by some of these pioneering organizations in several countries in Asia, Africa and Latin America is given in Supplement A, *Appropriate Technology Organizations in Developing Countries* (pp. 192–246).

Part Two

TECHNOLOGY CHOICE IN RICH COUNTRIES

3

TECHNOLOGY: THE CRITICAL CHOICE

The drastic effect which a change in technology can have on people and their jobs has been well known since the Industrial Revolution, but most people have taken it for granted that technology is somehow beyond human control and that its natural and inexorable development must be ever onwards and upwards: towards greater size, cost and complexity. Schumacher was the first to draw attention to the pervasive influence exerted by the choice of a technology and to the need to make technology choices available. He then went on to prove that it was possible to offer developing countries a range of technologies by launching the Intermediate Technology Development Group.

The proposition that technology choices exist, or can be created, is not on the face of things a revolutionary doctrine. Yet it is undeniably a revolutionary step to start putting economic power into the hands of people who have little or none. And this is precisely what is starting to happen now that the myth of 'technological determinism' has been exploded. We have seen, in the developing countries, the many types of initiatives which become possible when people in rural communities are furnished with the technologies of self-help. Over the same period, that is, during the past ten years, but especially since the publication of *Small is Beautiful* in 1973, a significant 'alternatives' movement has grown up in several of the most highly industrialized countries. This too owes much to the basic concepts of intermediate technology – smallness, simplicity, capital saving and non-violence – and to this movement we now turn.

Not surprisingly, the first people in rich countries to discern that they too needed something on the lines of intermediate technology were those living and working in the hinterlands of large metropolitan economies. As early as 1968 I attended a meeting sponsored by the Memorial University of Newfoundland, and met people from western Scotland, north Norway, Iceland, north and west Canada and other territories – or parts of them – which are the poor areas of rich countries. It is characteristic of such territories that they closely resemble colonies (which produce, as someone put it, what they do not consume, and consume what they do not produce); and the faster the metropolitan centre that controls them grows, the more rapidly they deteriorate. If they are to do more than merely survive with the aid of welfare payments, such communities need technologies appropriate to their resources and their lifestyles. That was, in fact, the chief conclusion of the Newfoundland meeting,[1] which was the first of a series of Canadian initiatives in the field of alternative technologies and lifestyles, of which, as we shall see later, two outstanding examples are now in evidence in Sudbury, Ontario, and Prince Edward Island.

This insistence that economics and technology must spring from local culture and not dominate it runs counter to the centralist trend in all societies. Fortunately, people still object to being made the objects of rationalized production, especially if their lives are controlled by some remote and authoritarian body. This is the real pressure underlying demands for economic and political self-determination that have emerged in Scotland, Wales, Brittany, the Basque country and elsewhere – and also why, incidentally, Tasmania was the first relatively poor area of a rich country to set up an appropriate technology organization. There are, of course, many more of these 'mini-economies' than meet the eye: only a few have the political and cultural cohesion sufficient to demand more self-determination and, in more extreme cases, political separation.

Its destructive impact on its own hinterland is only one of many self-defeating characteristics of modern industrial society. So it was a short step to the further question: what role could there be for alternative technologies in the rich countries themselves? The question could hardly have been more timely, because during the late 1960s and early 1970s many previously held assumptions about the goals of industrial societies, and the means of achieving them, began to fall apart. Questions began to be asked – and they continue to be asked with increasing urgency – about the role of industry in society, about environmental damage and resource depletion, and about the future of energy supplies. In all of this, technology was and remains a central issue. More precisely, it can be asked: if the urgent tasks of industrialized countries are to find ways of humanizing industry, protecting the physical environment and conserving natural resources, is not a new kind of technology required – smaller, less rapacious, capital and energy *saving*?

The damage done to men and women and their families by the growth of large-scale industries and the lifestyles they impose on society – boring, meaningless work, production and consumption seemingly geared to maximizing waste, urban congestion and squalor, and the diminishing power of most citizens to influence this economic and social environment – such criticisms have been levelled by the thoughtful and courageous throughout the growth of conventional industrialization. In the Western tradition, this line of criticism – the attack on the dehumanizing aspects of industry – runs (with variations) from the great religious leaders through Marx, Ruskin, Morris, Kropotkin, Tawney, Titmuss, Gandhi, Illich and Schumacher. It is also in the Western capitalist tradition for the power élites to listen to such men with interest and even respect, and then to ignore them in practice. But by the late 1960s the accumulating debits, in human terms alone, of the mass production, let-it-rip economy were becoming hard to ignore. The growing disaffection of young people, of various formerly quiescent and under-privileged groups in manufacturing and service industries, became evident, and continue to make themselves felt. These groups include women, ethnic minorities and community organizations.

The conflicts between the expected norms of behaviour of workers in the modern factory, and outside it, formed one of the themes of Richard Titmuss's brilliant *Essays on the Welfare State*,[2] published nearly twenty years ago: in terms of stability, certainty of status, and initiative, there is a fundamental clash between what is imposed on industrial workers within the factory, and what is expected of them outside it as husbands and wives, parents and citizens. The consequences are, of course, not only personal but also social, economic and political. After a lapse of two decades these issues have been taken up again and explored in detail by Fritz Schumacher, Hazel Henderson, James Robertson and others.[3]

The recent upsurge of interest in job enrichment and worker involvement is a concession to the fact that the content, the quality of work, is somehow important. At its worst it recognizes that the trend towards job impoverishment does not pay; at its best it acknowledges that work is the greatest formative influence on people. Whatever the motives, it accepts that technology can be damaging, and that a conscious effort is required to make it less damaging (or more constructive).

But is conventional technology in the form of large-scale, highly concentrated industry capable of providing enough employment of any kind at all, of whatever quality? The total dependence of the mass of industrial workers upon the vast, Kafkaesque, impersonal structures of modern industry is now being highlighted by growing unemployment throughout the industrial countries. A technology deliberately aimed at achieving production without people can hardly be expected, in the real world, to produce any other result. The fact is that we can no longer rely upon conventional large-scale industry to provide enough work – quite apart from creative or satisfying jobs – to meet the human and social needs for useful employment. For that we must develop alternative structures and alternative technologies.

The failure of large-scale corporations to create employment is illustrated by this evidence from the U.S.A: between 1969 and 1976, total employment in the U.S.A. went up by nearly 10 million – but only one per cent of this increase occurred in the 1,000 largest companies in the U.S.A. (the 1,000 firms listed by *Fortune* magazine each year).[4] Similar figures are not available for Britain, but since the concentration of industry into very large units is much the same in the two countries, the employment-creating, or rather unemployment-creating capacity of the large firms in the U.K. is unlikely to be radically different. Yet it is into such industries, rather than into any alternatives, that the U.K. government has pumped more than £3,000 million during the past five years.[5]

To the growing social tensions created by mass production there was added, during the early 1970s, a second wave of protest against the consequences of modern industrialization. This time the spokesmen were scientists and others concerned about environmental destruction. This movement is associated with the names of Barbara Ward, Rachel Carson and Barry Commoner, with the Club of Rome, *Limits to Growth*, *Blueprint for Survival*, the U.N. Environment Conference and the U.N. Environment Programme, and now increasingly with many excellent organizations such as Friends of the Earth, International Institute for Environment and Development, and the Worldwatch Institute. They represent a rapidly growing concern about the violence inherent in modern technology and industrialization towards both living things and the inanimate environment, a concern that continues to mount as the extent and variety and danger of pollution, and the rate of depletion of the earth's natural resources, become revealed. No 'hidden hand' of the market can avert dangers such as pollution and the environmental hazards bound up with the exploitation of nuclear energy, which to our peril is becoming an integral part of conventional economic growth. Care for humanity and the environment demand careful and informed judgments about the kind of technology we use and about the quality of economic development that is consistent with a sustainable future.

For many of us, the human and environmental damage wreaked by modern technology would by itself be enough to call for a drastic reappraisal of conventional notions about economic growth and how it can be achieved. Now the impending energy shortage leaves no room for doubt about this necessity. The sudden end – in October 1973 – of the era of cheap and abundant energy has dealt a crushing blow to the very basis of industrial economies – to the structures of agriculture, industry, urbanization and transport that the rich countries have built up, and which are based on cheap oil. Now, whatever else may be in doubt, it is beyond question that the need for a new approach to technology in the industrial countries has become not a matter of political expediency and intellectual speculation, but a condition of survival. The growing awareness of the damaging effects of our form of industrialization upon people, on the quality of life and the environment, started the process of making us think about alternative technologies: technologies that can express the humane application of knowledge – that are non-violent towards people and natural resources. The energy shortage has clinched the matter.

In reality, therefore, the needs of rich and poor countries are by no means so far apart as is often supposed.

The poor countries of the world need intermediate technologies in order to raise the mass of their population to a decent standard of life. There is manifestly no way of doing so by means of the capital and energy-intensive technologies which form the main currency of aid and development. The rich countries stand no less in need of a new technology, one which will enable them to arrive at sustainable lifestyles, based largely on low energy and renewable resources.

Invariably, when the subject of the need for intermediate technologies in rich countries is being discussed, someone will raise the objection: 'Intermediate technology can have no application here. It is a labour-intensive technology, and we are high labour-cost economies; in fact one of the chief reasons why capital-intensive methods have been developed is precisely because labour costs are so high. So to go for labour-intensive instead of capital-intensive methods is simply to step from the frying pan into the fire.'

On one of the occasions when this argument was raised, Schumacher responded as follows:

> When I am asked what science and technology should concentrate on, I reply that we need methods and equipment which are
>
> – cheap enough so that they are accessible to virtually everyone;
> – suitable for small-scale application; and
> – compatible with man's need for creativity.
>
> In the past it has been (and it still is) assumed in our modern and very Western discipline of economics, that, for instance, more capital automatically decreases the need for labour or more labour automatically decreases the need for capital. But we have also been assuming other things. Some of them are so wrong-headed as to be dangerous: for instance, the assumption that work could be made stupid and mechanical and yet productivity could be maintained or increased by increasing wages. Other assumptions we have made are so wrong-headed that I would describe them as wicked, for instance that human work is simply and completely a burden, in economic terms a factor

only on the 'cost' side which should be eliminated or reduced in any way possible. Yet the entire experience of mankind demonstrates clearly that useful work, adequately rewarded in some combination of material and non-material things, is a central need of human beings, even a basic yearning of the human spirit. You will notice that I did not say an *artist's* need for creativity, or an economist's or scientist's need for adumbration of an elegant model, or an engineer's or technologist's need for useful outlets for his creative powers: I said *man's* need, the need which lies within every one of us.

This is not a gratuitous flight into metaphysics; it is a hard fact of life which the economists and systems analysts and professional managers of a generation or two from now – assuming these vocations exist then under their present names and assumptions, which is questionable – will recognize in their everyday work as they do now, perhaps unconsciously, in their everyday lives. In the meantime, we have a very rough patch to get through. The kind of technology which we need in the industrialized West is very different from what we have now; yet we have barely even begun to imagine what a good deal of it will look like. We are hypnotized, perhaps paralysed, by the glitter and the size of modern industrialized technology to the point that when we think of alternatives we often can think only of long-haired people on a farm trying to re-create a mythical 1910 or 1880 or 1840.

Many of us think there is no relevance to industrialized nations of the world of the kind of thinking embodied in the concept of intermediate technology, because we equate this sort of technology with 'labour-intensive' and we think that with our high wage scales in the West the necessary amount of labour could not be paid. End of thought, end of exploration, end of effort.

Perhaps we may need the intellectual adrenalin which comes from the fact that what we are really talking about in the West is rising wages coupled with static or declining productivity, and the fact that this is just one of the areas in which the mechanical economic model – with, in this case, the assumption that higher wages evoke greater productive activity – is breaking down completely. In the United States they have coined the term 'stagflation', for the combination of inflation with stagnation which represents another facet of the bankruptcy of the mechanical economic model and essentially its ignoring or disregarding the way living human beings actually act in relation to one another in the world.

One way of thinking about the role of intermediate technologies in industrialized nations is to think of 'labour-intensive' and 'capital-intensive' as being a kind of Hegelian thesis-antithesis from which a new synthesis is emerging. In the context of the poorer countries, where the role and form of intermediate technologies are much further advanced than in industrialized countries, we often speak of 'skill-intensive' or sometimes 'thought-intensive'. The underlying assumption is the not merely useful but essential role played by individual thought and action, 'creativity' broadly considered. In fact, such creativity is, in our view, an essential part of the true 'economics' of development in any society: and we are the more impoverished because we have for so long ignored it.[6]

It may well be the case that this ignoring both of the individual capability for initiative, for creative and self-directed effort, and of the human need for a useful outlet for creative powers in work, has put us in the West in a comparatively more desperate position than are the poor countries. To go back to our original question, about the possibility of greater utilization in industrialized countries of intermediate technologies: we are partly handicapped by our high wages, but we are even more handicapped by our continuing obeisance to the labour-intensive versus capital-intensive dichotomy and the resulting assumption that we cannot 'afford' technologies focused more on the role of human effort and therefore must rely yet more on higher and higher capital investment.

Yet the capital which would be required seems to be in shorter supply than ever. In other words, as long as we continue to think in conventional terms we are finding that we cannot afford *either* the capital *or* the labour for new technologies, whether along currently accepted, high-capital-per-workplace lines or along intermediate, lower-capital-cost lines.

This labour-intensive/capital-intensive dichotomy – the argument that you can have either one or the other – which is presented to us as fact, is really nothing more than a series of assumptions, about people, work, capital formation, and technology, which are based on very shaky premises. Consider, for instance, the universal experience of the decline in the productivity of capital, the decrease in the profit or surplus of productive enterprise. This squeeze on surpluses is partly a result of a century or so of disregard for the dimensions and nature of inanimate, non-renewable resources. A heedless policy of treating this capital endowment as if it were income, and always going for the cheapest and most easily accessible resources first, is now confronting us with increasingly inaccessible and costly resources.

In part, too, it is owing to a similar disregard for the limits of tolerance of living nature: and sheer self-preservation now demands that a growing proportion of real resources must be turned to the avoidance of environmental destruction.

But most of all it is due to a disregard of the nature of human beings, and the primacy of those of their needs that accompany and are not subordinate to 'survival' and economic needs: the human needs for satisfying and creative work, a sense of belonging and co-operation and mutual respect, and other non-material rewards that everyone seeks but seldom finds in modern industrial society. The result is what Schumacher refers to in the quotation above, namely a combination of escalating wages *and* declining productivity, or 'stagflation'.

Now all this is a result of the way we have chosen to exploit non-renewable and renewable resources, and people; in other words the declining productivity of capital is largely a function of the technology we have developed and its damaging effects on resources, environment and people.

If there were no alternatives, our situation would be hopeless indeed. But there are: we can develop technologies that rely much less on non-renewable resources – that are capital-saving – and use renewable resources on a 'sustainable yield' basis; and devise technologies and lifestyles that rely less on huge, centralized agglomerations of capital, and more on the skill and creative initiatives of human beings.

We have hardly even started to explore the range of technology choices open to us, which could arise out of a mobilization of people's creative initiatives. It was to illustrate this point that Schumacher invented his 'law of the disappearing middle':

In technological development there is some very primitive

technology, let us call that Stage 1, then there are improvements, let us call that Stage 2, and then there is further
improvement, a higher level is reached, which we call Stage 3,
and Stage 2 drops out, virtually disappears. Then it moves to
Stage 4 and Stage 3 drops out; and so by the time you have
reached Stage 12, there is left only Stage 1 and Stage 12. That is
why we set up the Intermediate Technology Development
Group to fill in these missing stages, for the little people: for the
people who are helpless otherwise, who are reduced to the hoe
and the sickle because they cannot afford the combine
harvester. It is like in a bookshop, you can buy immediately the
most recent books or the classics, but any book that appeared in
1965 is unobtainable.[7]

There are, in short, many possible technologies between Stages
1 and 12, between the highly labour-intensive technologies at one
end of the scale and the highly capital-and-energy-intensive technologies at the other extreme.

The inescapable need to find alternative ways of producing the
goods and services we need in the industrialized countries has not
yet entered the consciousness of those who run the power structures in our societies, governments, industry and commerce, the
trade unions, the bureaucrats – or if it has, they are keeping quiet
about it. They continue to prescribe the mixture as before, they
continue to strive after economic growth by stimulating conventional investment by pouring public funds into large-scale
industries, and to support the proliferation of the most violent
technologies imaginable, including armaments, nuclear power and
chemical agriculture.

Fortunately for us, not everyone is prepared to join in the
process of keeping alive a system which is already in widespread
disarray. In all the industrialized countries there are more and
more people with courage and vision enough to work on alternatives, to give practical expression to the principles of smallness,
simplicity, capital-saving and non-violence. You can find their
groups and organizations in almost every department of human
activity – in agriculture and industry, in housing, distribution,
education and other social services. Although as yet relatively
little is heard of it, this development of alternatives is unquestionably one of the most important movements today, on both sides of

the Atlantic. It is not, of course, a unified or co-ordinated effort; indeed it is perhaps a movement only in the sense that it represents a multitude of ways in which the creative intelligence and power of individuals and groups can be constructively and effectively mobilized. But this, the discovery and mobilization of people's power, may be nothing less than the condition of survival for the formerly affluent societies of the West.

As Hazel Henderson says,

> In any period of cultural transition, the dominant organs of a society often increase their efforts to reassure the public, while their leaders privately express doubt and fear. This is not surprising, since it is precisely these institutions of government, business, academia, labour and religion, as well as their leaders, which are in decline and whose power is threatened and eroding. The information gathering and disseminating media, the statistics and the indicators are all geared to measuring society's wellbeing *in terms* of the wellbeing of these existing institutions. Therefore, the growing shoots of these societies go unmeasured and are overlooked and will remain insufficiently monitored and studied as possible new social models. We cannot afford to wait until the conceptual wreckage of industrialism is sifted and composted. We need to study the counter-economy at the same time that we are examining our now inappropriate statistics.[8]

The chapters which follow are meant as a contribution to making more widely known the work of those who are building the new social models and the counter-economy. Three of us – Bill Ellis, the founder of TRANET (Transnational Network for Appropriate Technology) and editor of its journal, my son John and I – worked together to compile these accounts of the salient features of the alternatives movement in Britain, the United States and Canada.

4

A GUIDE TO THE ALTERNATIVES MOVEMENT IN BRITAIN

Alternative Agriculture

The sustainable and ecologically non-violent alternative to conventional – that is, largely chemical – farming is biological husbandry.

Like that of other industrialized countries, Britain's agriculture and food production is very vulnerable to the growing oil shortage. While farming alone accounts for only about 4 per cent of the nation's use of primary energy, if we include the energy it takes to bring food to our tables, the figure rises dramatically to nearly 25 per cent of total energy use. For every unit of energy we consume in the form of food, we expend ten units of fossil fuel, mostly oil.[1]

The extent of this non-renewable energy dependence would, by itself, require major changes in the agriculture and food supply system, to make it more sustainable. But alone among the major industrial countries, Britain does not feed itself: about 50 per cent of our food is imported. In a world where food surpluses are likely to disappear quite soon, Britain's future food supplies are very far from secure.

This massive expenditure of energy is partly a result of the increase in farm mechanization and in the use of inorganic fertilizers, herbicides and pesticides. During the past fifty years the mechanical horsepower on British farms has risen sixteen-fold and the use of nitrogenous fertilizer, twenty-fold (while total production has only doubled). Mostly, however, the food supply system is responsible for this high degree of oil dependence. As Gerald Leach points out:

It has been the industrial-urban system, its values and economic forces, that has caused the food system to change so as to greatly increase its energy demands: double and triple packaging of foods; greater transport distances; the strident selling of novelty and convenience foods of ever-greater artificiality in order to get more 'value added' when nutritional intakes have virtually flattened off; the genuine desires of housewives to save time in food preparation; supermarket economics with their 'slice, wrap, freeze' practices in order to save labour . . . the list is almost endless.[2]

A sustainable, permanent agriculture and food supply system requires a shift away from chemical farming towards biological methods, and a much higher level of food self-sufficiency. The latter in turn implies that more people should be growing food, that more food is processed locally for local consumption, and that there is a shift from animal to vegetable protein consumption. Here is how Schumacher put it in his last presidential address to the Soil Association:

Let us see what would be the three musts, the three most desirable things about agricultural achievement in this country. First, we must have an absolute determination backed by work, not by statistics, to attain food self-sufficiency . . . Secondly, and I know that this will be received as a very tall, or impossible, order, this has to be achieved by biological instead of chemical means, because as more and more people see quite clearly, the chemicals, all based on fossil fuels, will simply not be available. Thirdly, I believe we should conceive the necessity that, in the course of the next few decades, the proportion of active population on the land will be enabled to rise from what is, currently, about 4 per cent, to something of the order of 15 per cent.[3]

The need to involve more people in food production arises not only because this is the only sustainable way of raising output per acre (which is the only important factor where food production is concerned – not output per man) but also because, if very few people live and work on the land, the urban population becomes convinced that the sole purpose of agriculture is to produce food as cheaply as possible. This is a delusion. Agriculture has several, very important, secondary tasks. One is to maintain an agreeable and healthy landscape. Another is to maintain genetic variety, now being threatened by monoculture. No less important is the maintenance of what Schumacher has called the 'water household' of nature, by agricultural practices that inhibit extremes of drought and flood, maintain the water table, and keep the water pure.[4]

That a sustainable agriculture and food supply system using biological methods is possible is now beyond reasonable doubt, thanks to the work of a number of voluntary organizations and dedicated individuals. (Government research establishments resolutely ignore biological husbandry.)

The principal organizations promoting biological husbandry are the Soil Association, the Organic Farmers and Growers co-operative, the Pye Research Institute, the Henry Doubleday Research Association, the International Institute of Biological Husbandry, and Emerson College.

The Soil Association, with 4,000 members, runs courses on organic methods as part of its educational activities and, over the years, has made notable contributions to the scientific and practical understanding of biological husbandry. Growers who conform to the Association's code of practice are licensed to use its trade mark as a guarantee of the quality of their products. The Association also runs a tree trust started by Schumacher to promote the planting of food-bearing trees, especially on land unsuitable for cultivation. A few years ago the Association launched a co-operative, Organic Farmers and Growers, which now has approximately 100 members with holdings ranging from nearly 4,000 acres to under one acre. The co-operative produces manures and other supplies for its members, and runs an advisory service. Many of its members in fact joined the co-operative in order to make a transition from chemical to biological farming. It also helps its members to get organically grown seeds. The chief activity of the O.F.G. is marketing, and it is becoming hard for the co-operative to keep up with the demand for organically grown cereals, vegetables and herbs.

The Henry Doubleday Research Association is the largest grouping of organic gardeners (rather than farmers) in Britain. It has nearly 7,000 members, twice as many as four years ago. The

Association has a two-acre experimental garden at Bocking in Essex, and generally more than 200 of its members are working on different experiments, with the Association acting as an information centre. It also runs a fruit and vegetable library, with the aim of keeping on the market varieties that would otherwise disappear.

A very valuable supporting service to non-chemical agriculture is now being provided by the International Institute of Biological Husbandry (I.I.B.H.), formed in 1975 to promote the scientific development of biological and organic agriculture as a viable alternative to conventional farming. One of the basic concepts of biological husbandry[5] is that there is an indivisible connection between the health of the soil and that of the crops grown on it and the animals and people who consume these crops. The basic principle is to 'feed the soil' and its biological life, rather than to present the crops with large amounts of soluble nutrients in the form of inorganic fertilizers.

The I.I.B.H. is building up a network of concerned scientists and agriculturalists throughout the world and has started to produce very carefully documented material on the scientific case for, and science of, biological husbandry. Its first publication, *Who Needs Inorganic Fertilizers Anyway?*,[6] disposes of the widespread assumption that it is only by using large quantities of inorganic fertilizers that enough food can be produced for the growing world population; and goes on to show that biological agriculture is capable of competing with chemical methods. (It also explains, very clearly and simply, what biological agriculture is and how it differs from chemical methods.)

Among the centres that offer comprehensive training in organic farming and gardening there is Emerson College, a training centre of the Rudolf Steiner movement for biodynamic agriculture, a specialized form of biological husbandry; and the International Institute of Biological Husbandry, at Country College, Well Hall, Alford, Lincolnshire. More specialized courses, for gardeners and smallholders, are run by the International Association of Organic Gardeners, and by the Soil Association; and in Wales at the Centre for Living and the Smallholders' Training Centre.

Recognizing that until organic methods are officially approved and accepted as orthodox training courses, they will not become widespread, several organizations have recently banded together[7] to form COMET (Combined Organic Movement for Education and Training). Its purpose is to provide and encourage education and training in organic methods. (At least two education authorities – Manchester and Dundee – have recently given awards to students to attend one-year full-time horticultural courses at the centre of the International Association of Organic Gardeners.) COMET now produces a newsletter giving details of organic courses throughout the country.

The Pye Research Institute is conducting practical experiments on the comparative performance of organic and chemical farming on its farm at Haughley in Suffolk. The farm is divided into three; one part is entirely organic, another is based on a mixture of organic and chemical methods, the third uses wholly chemical methods. The Institute has data on comparative results going back over twenty years. It has also done valuable research on nutritional and other aspects of food grown organically.

A very promising research programme has recently been started at the University of Reading, under Professor Colin Spedding. Its starting point is that very little attention has been given to the needs and potential resources of the smallholder, whether full- or part-time. The intention is to develop an appropriate biology for the smallholder. Crops on large farms, for instance, are designed to feed machines; plant breeding programmes aim to develop plants that can be mechanically cropped in the space of one or two days. But the smallholders' need is for plants that can be cropped over a long period. Similarly, most animal breeding is done with the large farm in mind. Professor Spedding aims to identify plants and animals – and a supporting technology – that will specifically suit the smallholder. This programme also includes research on food production in urban areas.

The upsurge of interest in self-sufficiency and organic gardening in Britain is not restricted to rural-based organizations.[8] One of the most imaginative and successful programmes launched by Inter-Action, a group concerned with community participation and self-help, is City Farms. The first City Farm was started in 1972 in north London, on the site of an old timber yard and a British Rail shunting depot. This was transformed into allotments, a gardening club and a barnyard, where local children helped to rear sheep, pigs, chickens and goats. The idea caught on rapidly and, within a few years, similar farms had been started on several other London sites and in Birmingham, Bristol, Liverpool, Nottingham, Swindon, Newcastle and Glasgow; and thirteen other sites were under negotiation by mid-1978. These City Farms are not, of course, concerned with the commercial production of food. They are places where:

> the house cow or the backyard pig can become the responsibility of a small group of adults or children for the benefit of their neighbourhood . . . Inter-Action encourages the local management of City Farm Projects in the belief that the current physical decay and social need prevalent in cities are best tackled by those who have their roots, their families and their futures in the neighbourhood.

Inter-Action has now set up a City Farm Advisory Service to advise local groups on how to set up, organize, finance and manage City Farms for themselves.[9]

Lucas Aerospace Workers' Corporate Plan and Other Workers' Plans

Within the manufacturing industry in the U.K. there is a growing realization among the workforce that something has gone badly wrong. The 'white hot technological revolution' has destroyed jobs and brought about more hazardous, fragmented and de-skilled work. Job redesign and automation have meant a shake-out of conventionally skilled craftsmen, and declining job security for the unskilled and semi-skilled.[10]

Training and re-training schemes for apprentices and skilled operators have not matched in either size or quality the pace of technological change. Even now, few companies are willing to establish training programmes without substantial government assistance.[11]

The markets for U.K. armaments, motor cars, all U.K. engineering equipment and aerospace machinery have become very insecure. Continued dependence on the consumption of massive quantities of oil and other natural resources has made U.K. industry exceptionally vulnerable to even minor changes in the demand for its products, or the supply of its raw materials. This has left the industrial workforce and the trades unions exposed to the threat of sudden redundancy, brought about by circumstances which they have no power to influence.

The dominance of large, multinational companies in the national economy has increased and so, too, has their tendency to shift work and investment outside the U.K. This has happened despite the generous grants and loans made available to private industry (and mainly large firms) under the 1972 Industry Act and the government schemes for tax relief on stock appreciation and deferred corporation tax. Public accountability for these disbursements from the public purse is negligible.

These characteristics of the manufacturing economy are viewed with alarm by the workforce within industry. Yet it appears that the government and the trades union movement either feel powerless to halt the diminution of the manufacturing base, or are indifferent to the loss of output, skills, manpower and variety within it. For quite dissimilar reasons, neither institution has even questioned the sole right of industrial managements to determine the nature of manufacturing production. No government, in the shape of a strongly conventional, conservative-minded Department of Industry, considers that it should influence such decisions. The trades unions, for their part, show no signs of departing from their orthodox line of fighting for better pay and work conditions for their members, and for eventual public ownership.

Instead, the determination to control the composition and end-use of production within private companies has emerged from the workers in individual firms. The first group of workers to take the initiative were employees of Lucas Aerospace Ltd. In their campaign to fight for the right to work on socially useful products[12] they have emphasized that the products of industry must:

be accessible and useful to everyone in the community and not confined to fulfilling the needs of the select few;

take maximum advantage of the existing skills within a company and develop them to the advantage of the workforce and the community;

be capable of being made and used in ways that do not impair the health and safety of the workforce or the community at large; and

make minimum demands upon natural resources and improve the quality of the environment.

In order to begin working towards these ends, in 1968 the shop-stewards in Lucas Aerospace formed the Lucas Aerospace Combine Shop Stewards Committee (L.A.C.S.S.C.) to co-ordinate trade union policy and the negotiations for better wages, conditions and job security in the seventeen Lucas Aerospace sites in the U.K.[13]

The Combine Committee is unique in the British trades union movement. It brings together people from the highest level technologies and the skilled and semi-skilled workers from the shop floor. For these elected shop steward representatives it provides a forum in which to discuss topics and strategies relating to all the company's factories.[14]

This adventurous step taken by the shop stewards reflects the growing awareness among working people that conventional trades union structures, founded on geographical divisions and built up on a craft basis, are unequal to the task of challenging the organizations of production and employment imposed by multinational firms.[15]

In 1974 the Combine set up a Science and Technology Advisory Service to anticipate the problems associated with new technologies and processes.[16] Assessments were made by the workforce with occasional help from outside scientists and technologists. In particular, the Combine wanted advance warning of the likely human consequences – fragmentation of skills, increased work rates, poorer job security and unfamiliar hazards – of new production methods and management techniques.[17]

However, in the same year it became clear that this limited defensive tactic could not meet the much wider problems facing the workforce.

The effect of the 1973–4 oil price rises coupled with the rationalization policy pursued by Lucas Aerospace, which had reduced the workforce by 4,000 between 1970 and 1974, were only two of a number of factors which convinced the Combine that the company's existing policy, its product range, spread of markets and the type of production techniques likely to be produced by the management were not going to guarantee employment or reverse the trend toward the brutalization and de-skilling of work. Impending large-scale redundancy within Lucas Aerospace was a very real and serious threat. This the Combine was prepared to oppose openly. But they were also determined to begin tackling the contradictions that they recognized within the industrial structure of the U.K.: contradictions that they felt tended to be overlooked or ignored by trades unionists, economists and other commentators in their search for the reasons underlying redundancy and the debasing of skills.

First, 'there is the appalling gap which now exists between that which technology could provide for society and that which it actually does provide.'[18] Industry can and does provide marvels of technological sophistication such as Concorde. What it is apparently incapable of providing, however, are inexpensive, simple heating systems for the elderly; or a sufficient number of haemodialysis machines to prevent the annual death of 3,000 kidney patients in the U.K.

Second, the Combine pointed to the glaring irrationality of a society which denies to one and a half million people any creative outlet for their skills, enthusiasm and ingenuity, while bemoaning its inability to produce more, and especially to maintain essential services.

The third point, as Mike Cooley, a member of the Combine and past President of A.U.E.W.-TASS, has put it, 'is the myth that computerization, automation and the use of robotic equipment will automatically free human beings from soul-destroying, back-breaking tasks and free them to engage in more creative work. The perception of my members and that of millions of workers in the industrialized nations is that in most instances the reverse is actually the case.'[19]

The fourth contradiction the Combine identified was that public hostility towards the application of science and technology was generally directed towards the individual scientists and technologists. Yet, those commonly held responsible for the degrading consequences of applied science and technology in dehumanizing work and consuming large amounts of non-replaceable natural resources are, in practice, given no opportunity to determine the nature and direction of their work, which is defined for them by their employers.

The Combine recognized that the impending redundancies within Lucas Aerospace and the wider problems of the role of large-scale industry in society needed to be tackled together if they were to be tackled effectively, and that the response would need to be more radical than the traditional but increasingly ineffective trades union strategies of fighting for the right to work through strike action, site occupations and 'work-ins'.

Tony Benn, the Minister for Industry, met the Combine Committee in November 1974 and suggested that they might consider alternative products and 'intermediate technologies' that could be produced by the company in the event of a recession in the aerospace industry. The Combine decided to prepare an overall corporate plan, including a comprehensive set of proposals for the alternative technological development of Lucas Aerospace.[20]

This ambitious task involved a great deal of painstaking work by the shop stewards, technicians, scientists and production workers and occupied most of their spare time over a period of fifteen months. Initially the Combine asked for ideas and suggestions from 180 leading authorities, universities, institutions and trades unions, which in the past had expressed concern about the dehumanizing qualities of modern technology and had suggested a need for socially responsible, 'alternative' technology.[21] There was practically no useful response.

The Combine then turned to their own members for ideas. A detailed survey was made at each factory to elicit ideas and stimulate the imagination of the workforce. At each site plans were devised, and production techniques identified, for specific alternative products based on the local knowledge of available skills and equipment, existing products, site layout and services. Combine members also circulated feasibility studies, technical reports and reviews of current work in the fields of energy production, transport, medical technology and economics to back up the discussion of ideas for new products. They also sent out their own bi-monthly illustrated newspaper, *Combine News*. In many instances outside experts offered their help with new product lines and processes. The shop stewards also contacted potential customers, environmental organizations, local colleges and community groups and trade councils for their views on what products should be included. Towards the end of 1975 the

Combine began assembling the material they had collected into a coherent plan. [22]

After several drafts had been circulated for discussion and reappraisal, the final version was made public in January 1976. [23] It consisted of six documents, each of about 200 pages, containing technical details, engineering drawings, costings and supporting economic data for some 150 new products. Also included were strategies for the fundamental reorganization of production practices.

The Plan has become the central feature of the Combine's campaign for the right to work on socially useful products. First, it is seen as the basis for preserving jobs and skills at Lucas Aerospace. Second, it puts forward a series of alternative products that can be manufactured in ways that do not dehumanize or fragment the skills of the workforce. [24]

In the section dealing with alternative energy technologies the Plan advances an imaginative range of ideas for gaseous hydrogen fuel cells, heat pumps, solar collecting units and small wind electric machines which cater for energy needs of communities rather than individual households. Lucas Aerospace already makes kidney dialysis machines and heart pacemakers. The Plan proposes that this work should be stepped up and extended to include further items such as simple, portable life-support systems for the victims of coronary failure and mechanical 'sight' mechanisms for the blind. The workforce at the Wolverhampton plant received wide acclaim for their 'Hobcart' design, a simple self-propelled vehicle for children with spinal diseases. [25]

Several of the proposals are for intricate telechiric gear for use in the hazardous professions of deep-sea diving, oil rig maintenance, fire fighting and mining. A number of related developments are also identified: the collection of ocean bed mineral-bearing nodules, and submarine agriculture.

The section on transport systems contains a variety of ideas. The hybrid diesel-petrol power pack, for example, uses a small internal combustion engine running at constant optimum revs to charge batteries (whenever the vehicle idles or decelerates) which are also used to provide motive power through an electric motor. Prototype tests (in conjunction with Queen Mary College, University of London) have shown that the pack dramatically lowers fuel consumption, toxic emissions and noise levels.

Another proposal is for a road-rail vehicle that can be driven as an ordinary coach on the roads and transfer directly on to a rail network. Studies on this vehicle have shown that by using pneumatic tyres and with a smaller, lighter frame than normal rolling stock it can cope with sharper track curves and steeper gradients (up to 1 in 6). These qualities make it much simpler and cheaper to lay track for the vehicle because the massive earthworks necessary for conventional railway systems are not required.

The Combine shop stewards point out in the section on job redesign that nothing will be gained from a campaign to work on environmentally sound and socially useful technologies if their manufacture perpetuates work conditions that are hazardous, polluting and mind destroying. They are also eager to do away with the conventional divisions between 'professional' management and shop-floor workers.

The organizational arrangements the Combine proposes are integrated production teams, where the semi-skilled and skilled manual staff and the high-level technical staff work closely together. Extensive re-training and education programmes are proposed at all levels in the workforce to extend and develop existing skills, and these would break down the barriers between manual and technical staff and help the workforce meet future technological challenges. The aim of these organizational and training proposals is to prepare all members of the workforce for the joint tasks of controlling production processes and the goals and priorities of production.

When the Plan was made public in 1976 it was widely acclaimed, not only in the U.K., but also in Sweden, France, Germany and the U.S.A. It was heralded by the *Financial Times*, the *Guardian*, *New Scientist*, the *Engineer*, *Industrial Management* and *Manpower* as a highly significant event for the future development of British industry in the fields of management and technology choice. Especial praise was reserved for the detailed technical work in the Plan.[26]

The shop stewards emphasise the principle of collective responsibility in all their work. But they have no wish to take over the running of the company or turn it into a workers' co-operative. Their purpose is to demonstrate that 'workers are prepared to press for the right to work on products which actually help to solve human problems rather than create them.'[27]

In line with these views the Combine was aiming for a phased introduction of the Plan. They did not visualize the company suddenly curtailing its deep involvement in the aerospace industry – a sector they recognized as important to the technological and economic base of the U.K. The industry was in decline. They proposed the gradual adaptation of spare capacity in the company to the manufacture of alternative products. Provision for a gentle easing in of the Plan also accounted for the time it would take the workforce and the unions to adjust to the organizational changes.[28]

In April 1976 the company published their response to the Plan. They rejected it and refused to consider any of the proposals it contained. Neither did they want to work with the Combine Committee.[29]

The Combine was shocked at the total rebuff of the Plan. It surprised many industrial commentators too, for they had expected a more considered response. A number of the leading scientific and managerial journals – including *The Economist*, *New Scientist*, *Industrial Management* and the *Engineer*, journals not particularly noted for their radical outlook – criticized the management for their negative attitude.[30] Management's brusque rejection, however, served to invigorate support for the Plan.[31]

The Combine continued trying to persuade the company to adopt the Plan. They approached the T.U.C. and the Department of Industry (DoI) for help to overcome the management's refusal to meet with the authors of the Plan. Official trades union backing was also sought.

The chronicle of exchanges between the Combine and the DoI, T.U.C., the management and trades unions makes depressing reading. It is characterized by evasion and equivocation on the part of each of the official bodies with which the Combine has come into contact.[32]

So, in February 1978 the Combine, along with the North East London Polytechnic, set up the Centre for Alternative Industrial and Technological Systems (CAITS), based at the Polytechnic. The Centre undertakes technical and theoretical development work for workers' corporate plans (with particular reference to the Lucas Plan) and local co-operatives. Technical projects include prototype work on the hybrid power pack and the road-rail vehicle, plus a number of others in the field of engineering, for example, fire escape apparatus for high-rise flats. In all projects, CAITS works closely with Lucas Aerospace workers.

CAITS has become the main centre disseminating information and background papers on workers' corporate plans.[33] (The demand for information on the Lucas Plan and the corporate plan principle has been overwhelming. Requests have come in from all over the world – from as far afield as Australia and China.) Three groups have been formed to investigate particular aspects of corporate planning. The Medical Group is exploring new design criteria for medical equipment, the potential for import substitution and new medical systems for developing countries. The Economics and Communications Groups are working along similar lines.

At the national level, CAITS has established close links with the labour movement and many leading academics. Locally it is helping to set up workers' co-operatives (studying housing and insulation) in London's dockland and works with local Co-operative Development Agencies and trades councils. The Centre has recently begun to publish a quarterly bulletin giving information about workers' plans and the lobbying campaign on their behalf. CAITS is also seeking to set up similar centres in other parts of the country, and a Bradford CAITS has recently started work in conjunction with the University of Bradford.

The greatest test for the Combine came in March 1978. Lucas announced closures of factories in Liverpool, Bradford and Coventry, making 2,000 workers redundant. On this occasion the DoI intervened. They provided the company with an £8 million grant towards new factories in Liverpool and Bradford. This deal did not mention the Plan and would have meant 1,500 redundancies.

The Combine formed an alliance with the Confederation of Shipbuilding and Engineering Unions (C.S.E.U.) which is officially recognized by the company. Each party in this alliance held deep misgivings about the aim and motivations of the other. But it was essential to take some action to prevent the loss of jobs. They worked together to prepare alternative proposals for the two sites.[34]

These were published in January 1979 as a mini-plan entitled

Lucas Aerospace: Turning Industrial Decline into Extension.[35] The report is deeply critical of management moves to shift work from the U.K. to Italy, Germany and France. It also criticizes what it sees as the unnaturally intimate relationship between the DoI and Lucas Industries. The report estimates that in deferred taxes and government grants the company has benefited by nearly £200 million. This has not prevented them from threatening to destroy 2,000 jobs.

In the rest of the report the shop stewards present a thorough analysis of the alternative product options at each site. They identify a variety of products: some for immediate production (haemodialysis machines), others with eighteen to twenty-four months lead times (gas turbines, fluidized-bed boilers, gas powered heat pumps), and proposals for future research and development work incorporating the facilities of several sites.

Following publication of the report, the company reached an agreement with all concerned. There would be no compulsory redundancies at Bradford and Liverpool for two years. A working party of company representatives and trades union officials would examine selected proposals in the report. The company agreed to make its 'best endeavours' to see that the commercially viable products recommended by the working party were manufactured.[36]

The company is still a long way from accepting the principle of socially useful production. But at long last it has been obliged to discuss diversification and the Plan proposals with the shop stewards.[37]

There is a hard task ahead for the Combine to prevent future redundancy threats and, at the same time, to press the company into alternative product diversification. Even so, they have already scored a number of successes. They have gained a widespread and very favourable public image for their almost visionary approach to industry and labour relations. (They were nominated for the 1979 Nobel Peace Prize.) More importantly for the workforce, since 1974 they have prevented any redundancies within the company. Their campaign has demonstrated that working people have the ability and the determination to control how and for whom industry is managed. One of the major accomplishments of the Plan has been to show that entrepreneurial talent is not limited to the few individuals who currently organize industry. The flair for discovering new ideas, new processes and new outlets for production extends throughout the workforce.

Conventional management, the Lucas workforce argue, is simply a command relationship, 'a bad habit inherited from the army and the church', as one worker put it. It is not a pre-ordained skill. It can be reshaped to suit the changing perceptions of the workforce and the community as to their needs.

The illuminating example of the workers at Lucas Aerospace has been followed by workers in other industries.[38] Threatened by factory closures in September 1976 at the two Manchester plants of Ernest Scraggs (a textile machine manufacturing firm), the 200 workers involved were prompted to fight their management's decision by drawing up proposals for alternative products. A Corporate Planning Committee held talks with L.A.C.S.S.C. Ideas considered were diversification into the machine tool industry, health and safety equipment and solar desalination equipment. Their efforts came too late, however, and both factories were closed.

Shop stewards at Chrysler Ltd have been considering ways to explore the feasibility of new, socially useful products in the light of 'widespread ecological and environmental criticism of the petrol driven car as a socially irresponsible form of transport'.[39] In a recent report, *A Workers' Enquiry into the Motor Industry*,[40] trades unionists from a number of motor firms urge workers to set up an industry-wide organization to define and campaign for alternative strategies to remedy the critical condition of the British motor industry. It suggests a number of possible products: hybrid propulsion vehicles, utility vehicles for the Third World and new mass transit systems.[41] The report is based on interviews conducted at a large number of factories.

Workers in the power engineering industry have formed a Power Engineering Combine Committee (covering such giants in the industry as C. A. Parsons and G.E.C.). They have put forward a variety of ideas for alternative developments in power engineering. Amongst these are proposals for the recommissioning of small (urban) power stations using advanced fluidized-bed combustion of coal to create combined heat and power units. These units would also provide district heating. Medium and large-scale aerogenerators for district electricity production or compression and storage of gases are suggested (with a view to production for the

Third World). Solar collectors, panels and cells have also been proposed. (Their campaign to preserve jobs in the industry is being supported and promoted by the Socialist Environmental Resources Association – SERA – see p. 272.)

The Vickers National Combine Committee of Shop Stewards have been developing proposals for diversification out of armaments production. Working closely with Mary Kaldor of the University of Sussex, they have produced two detailed reports: *Building a Chieftain Tank and the Alternative* and *Alternative Employment for Naval Shipbuilding Workers*.[42] The Vickers campaign has focused on maintaining employment at the Elswick and Scotswood sites in the north-east. At the Scotswood site, threatened with closure, the workforce, local trades unions and the local authority have been pressing Vickers to accept government assistance and modernize the plant to produce alternative products.[43] The main ideas proposed are for agricultural systems for developing countries (pumps, earthmoving equipment) and car presses. Other proposals are for heat pumps, fluidized-bed boilers, domestic refuse and metal recycling plants, mining equipment and machinery to improve canal systems. A number of sea-based technologies have been suggested – wave and tidal power generators, submersibles for fire fighting on oil rigs, deep-sea mining and marine agriculture and ocean-going tub-barge systems.

Workers at a number of other companies closely involved in armaments production have begun, mostly at an informal level, to consider alternatives. Rolls-Royce and British Aircraft Corporation workers have come up with ideas for machine tools, agricultural equipment, energy and transport systems. An alternative plan has also been drawn up by workers at the Dunlop plant at Speke, Liverpool. They have produced proposals for a domestic unit for recycling car oil, various small-scale medical devices, a rubber recycling process, and a bicycle which can be converted into a heavy-load-bearing tricycle.[44]

The principle of socially useful production is attracting a growing number of adherents within the official trades union movement. The Transport and General Workers' Union (T.G.W.U.) has published a document fully supporting the idea that the workforce should be allowed to develop schemes for alternative production in the event of redundancy threats.[45] A.U.E.W.-TASS have raised the question of alternative products for the shipbuilding, aerospace and computer industries. The Electrical, Electronic, Telecommunication and Plumbing Union (E.E.T.P.U.) has also called for re-training programmes and the introduction of new products in factories threatened with redundancy.

Similarly, in the U.S.A. trades unions and environmental pressure groups are beginning to criticize massive armaments expenditure and the very low birth-rate of jobs in large industry. The United Auto Workers and A.F.S.C.M.E.[46] are two large unions campaigning for full employment policies, emphasizing that more jobs can be created by a switch from military to civilian-oriented projects in mass transit systems, environmental protection and alternative energy technologies. These campaigns are being backed up by organizations such as Americans for a Working Economy and Environmentalists for Full Employment. In other countries such as Norway, Sweden and Germany trades union movements are also beginning to broaden their traditional campaigns to include consideration of who decides what and why certain products are made.

Co-operatives, Old-Style and New

Since the early 1970s there has been an upsurge of interest in small-scale enterprises, particularly co-operatively run firms, and of support and advisory groups set up specifically to help and promote co-operative and community self-help activities.

The reasons for this include heavy and growing unemployment, the desire to work in smaller units, to have genuine job satisfaction and to have some influence on the choice of products manufactured. (The only concession that government has made to the demand for greater industrial democracy was the Bullock report.[47] Its recommendations go nowhere near meeting the need. The report's proposals for industrial democracy, by setting the size of the workforce to be consulted at 2,000 on one site, excluded people working in two-thirds of private industry from having any say at their workplace.)

It is now recognized that the U.K. is sadly deficient in the co-operative organizations in industry, agriculture and transport that have proved so successful in France, northern Spain, Italy, Yugoslavia, Denmark, the U.S.A. and Canada.[48] Credit unions[49] and co-operatives serving and run by students and the handicapped are all well established in those countries.

The recent growth of small co-operatives in the U.K. has been associated with the creation of new structures of co-operative organization. The established co-operative movement in Britain has only very recently become involved with the new small-scale co-operative sector. From its formal origins in 1844 it has developed more with the interests of consumers than producers in mind. Although a producer wing of the early co-operative movement flourished up until the early 1900s, it declined steadily thereafter and there are now only seventeen of these old-style producer co-operatives in existence.

The failure of the producer co-operatives to establish a position within the growing co-operative economy of the nineteenth and twentieth centuries is, at least in part, due to the implacable hostility they attracted from the socialist intellectuals of the day. Sydney and Beatrice Webb conducted a life-long campaign vilifying producer co-operation and were strident advocates of consumer co-operatives and collective bargaining. They exerted a powerful influence on the policies of the early Labour Party and largely determined its preference for nationalization to democratic co-operative ownership. The critical and negative attitude of the Webbs, the trades unions, the consumer co-operatives and the Labour Party combined to stifle producer co-operation. Eighty years on, there are precious few signs of any mellowing of this attitude within the Labour Party or the trade union movement.

As the consumer co-operative movement expanded, it gave rise to the Co-operative Union (the national body representing and advising the movement in England, Scotland, Wales and Ireland), the Co-operative Party (a political organization now allied with the Labour Party), several specialist trade advisory services and two financial institutions: the Co-operative Bank and the Co-operative Insurance Society.

The seventeen old-style producer co-operatives, all formed before 1950, together employ over 1,700 people. They are all affiliated to the Co-operative Union, and nine of them belong to the Co-operative Productive Federation (C.P.F.), the only organization that specifically represents and co-ordinates the interests of these producer co-operatives. Most of the member firms of the C.P.F. are based in and around Leicester, engaged in the footwear, clothing and printing trades. Numbered amongst them are such long-established and very successful firms as Equity Shoes Ltd (founded in 1886) and Walsall Locks (founded in 1874).

Political hostility towards the early producer co-operatives was certainly a serious impediment to their progress. However this was not the only reason for their decline. They were hampered to some extent by the structure of co-operative ownership they adopted. In essence, the ownership of these co-operatives was (and still is) vested in the members of the co-operative who held shares in it. These included some or all of the workforce, retired workers or their families, outside sympathizers and institutions such as other co-operatives and trades unions. This meant that control of the co-operative was not primarily rooted in its workforce, some of whom may not even have been members. The value of the shares in the co-operative increased as its assets multiplied. In a few instances this led the members of the co-operative to close it down, sell the assets and distribute the proceeds amongst themselves.

The limitations of this co-operative model and the somewhat complicated procedures necessary to set one up, may have deterred people from following their lead.[50] What is certain is that, even today, very few people know of the existence of these successful and thriving co-operatives or of the model rules available to would-be co-operators from the C.P.F. The registration procedure to form a co-operative was made somewhat easier in the Industrial and Provident Societies Act 1965, which brought together all previous legislation over the past 100 years governing co-operatives. Even so, few of the small independent collectives and co-operatively run enterprises which were started up in the 1960s were registered as legal co-operatives. Nearly all of them remained informal and loosely constituted groups serving the 'alternative' sector – wholefood shops and restaurants, workshops and so on.

Factory closures during the early 1970s led to a series of attempts by redundant workers to organize workers' co-operatives in an effort to preserve jobs. Few of these got further than a statement of intent. Fakenham Enterprises was an exception and was set up with financial support from the Scott Bader Commonwealth (see p. 108). It lasted, fitfully, from 1972 to 1977. However, the most publicized of the defensive co-operatives were the three set up in 1974 and 1975: the *Scottish Daily News*, Kirkby Manufacturing and Engineering Ltd (K.M.E.) and Triumph Motorcycles (Meriden) Ltd.[51] All three were set up as workers' co-operatives following moves made by the owners to close down unprofitable privately owned firms, and received financial assistance from the government. In each case the decision to back them was taken principally with the view to saving jobs. The decisions were forced through at Cabinet level by Tony Benn, the Minister for Industry, in spite of the vigorous opposition of the Civil Service and the Industrial Development Advisory Board. The press, too, was hostile to workers' takeovers being financed from the public purse, and even the trades union movement viewed them with disapproval.

By 1980 only Meriden was still trading. The *Scottish Daily News* lasted only seven months and folded in November 1974. Fierce internal disputes, combined with a vague editorial policy, the boycott of the major advertising agencies and a weak market, contributed to its rapid extinction.

K.M.E. was set up in 1975 with a £3·9 million government grant. From the outset it faced problems. The original grant requested by the co-operative organizers was £6·5 million, a figure judged essential to meet the basic establishment costs of the enterprise. The grant they received was £850,000 less than was needed for the first year alone. Their problems were made worse by an inadequate management structure, an unlikely combination of products – domestic appliances and soft drinks, and the inflated price at which they were obliged to buy the factory buildings. None the less, the co-operative managed remarkable increases in output by introducing new work practices.

By 1978 the co-operative had received a further £1·5 million in government aid. It claimed 10 per cent of the U.K. radiator market and held orders worth £3 million. Armed with these figures and a favourable management consultant's report, K.M.E. approached the DoI for a grant to modernize the plant and lay off some workers. The DoI was unsympathetic to workers' co-operatives and refused.

Discussions with a private company interested in taking over the factory and some of the workers proved fruitless. In a final attempt to save the co-operative and the jobs, K.M.E. requested a grant of £5·8 million (later reduced to £4·5 million) to re-equip and make over 200 workers redundant. Again, the DoI rejected the application and in March 1979 K.M.E. went into voluntary liquidation with the loss of 720 jobs.

By mid-1979 the Meriden motorcycle co-operative was the only survivor of the 'Benn Co-ops'. It was started in March 1975 with a £750,000 grant and a government loan of £4·2 million. Despite initial marketing and management problems, its prospects looked promising. By 1978, output had risen by nearly 50 per cent, and industrial relations had improved beyond recognition.[52] A new, unpaid managing director started in January 1979, Geoffrey Robinson M.P., ex-managing director of Jaguar Cars Ltd, who had been instrumental in starting the co-operative. Meriden began looking for new markets in Europe, Australia and the Middle East to reduce its reliance on the U.S. market.

Meriden was given a new lease of life towards the end of 1980, when the government announced that it was writing off the £9 million loans and credits which had accumulated, and which had become an intolerable burden on the company. Meriden now have excellent prospects of prospering as a small company making a very high quality motorcycle, at the rate of between sixty and a hundred machines a week.

Criticism of the Labour government's aid to the worker co-operatives has focused on the 'waste' of public money involved. Unemployment rates in Glasgow and on Merseyside were, and still are, very high. The likelihood of redundant, specialized industrial workers finding comparable work, if it existed at all, was slim. The government would have been obliged to pay out substantial unemployment and social security benefits and to forego any tax income. K.M.E. has estimated that altogether these would have made the government poorer by £10 million if their workforce had been unemployed for four years.[53] Indeed, it has been calculated that the total cost to the government, in social security payments and lost revenue, of keeping a man in long-term unemployment is 70–90 per cent of his income in employment.[54] There is as much political significance in providing government aid to private industry or British Leyland as there is in supporting workers' co-operatives. Yet it would be a futile search to find the same vigour of press and bureaucratic condemnation of the 'Benn Co-ops' applied to decisions granting private industry large subventions from the public purse. There is no evidence to suggest that the workers in the co-operatives saw themselves in the vanguard of political change. Their sole concern was to preserve jobs and skills, and they readily accepted large cuts in wages to work in the co-operatives. At the *Scottish Daily News* the 600 workers invested £200,000 of their own money to help get the paper started. The failure of Fakenham Enterprises and two out of three of the 'Benn Co-ops' indicates very little about the merits or demerits of co-operatives in general. It merely reinforces the contention that 'an intractable situation of commercial difficulty is not transformed by change to co-operative ownership.'[55]

The collapse of K.M.E and the *Scottish Daily News* and the problems of Meriden illustrate that commitment and sacrifice by the workforce is not enough to ensure success. There is simply no substitute for management skills, and this expertise in the day-to-day running of a co-operative cannot be gained quickly.[56] Neither can the co-operative members develop the cohesion necessary to function as a business enterprise if they are subject to external pressures compelling them to establish the co-operative in great haste. Yet these were precisely the circumstances in which the 'new worker co-operatives' were formed. In this light, the achievements of Meriden and K.M.E. are very considerable.

Since the early 1970s, the growth of workers' co-operatives in the U.K. has been associated with the Industrial Common Ownership Movement (ICOM), which represents the main alternative structure to private or state ownership in industry.[57] The name ICOM was adopted in 1971 by the Society for Democratic Integration in Industry (DEMINTRY), formed in 1958 by Ernest Bader and Harold Farmer. Ernest Bader took the pioneering step in 1951 by founding the Scott Bader Commonwealth.[58] This was the model of industrial common ownership subsequently followed by a number of firms which converted from private to common ownership – Michael Jones Community Ltd, Trylon Community Ltd, Airflow Developments Ltd and the Bewley Community Ltd (best known for Bewley's Cafés in Dublin). Two firms were started as common ownership enterprises – Rowen Community Ltd and Sunderlandia Ltd.

ICOM, which has no party political allegiance, brings together new and established common ownership enterprises and institutions and organizations interested in this form of industrial ownership. It offers detailed legal, accounting and management advice to workers' co-operatives, common ownership firms and to companies wishing to convert to common ownership. In 1973 ICOM set up Industrial Common Ownership Finance Ltd (ICOF) as a separate non-profit making company to administer a revolving loan fund for established and prospective common ownerships.

The rapid growth in the number of co-operatives within ICOM took place after January 1976, when ICOM adopted a set of model rules for the formation of co-operatives.[59] These are a much simplified and up-dated version of the rules governing the old-established producer co-operatives. They limit the ownership of shares in a co-operative to the people who work in it and restrict the shareholding to one share per member. They also prevent the members from closing down the co-operative in order to enjoy a distribution of the assets. It is stipulated that whatever surplus remains upon dissolution of the co-operative must be given to another common ownership company or to a charity. These rules do not imply that ICOM has discarded the wider social objectives of co-operative enterprise. It has merely tried to avoid some of the inherent structural ownership difficulties that beset the old-style producer co-operatives.[60]

In November 1976 ICOM received a major boost when the Industrial Common Ownership Act became law. This made available, over a five year period, £250,000 to ICOF and £150,000 to 'relevant bodies' under the Act. These included ICOM (which receives the bulk of the grant), the Scottish Co-operative Development Committee and the Co-operative Productive Federation.

As a result of the availability of model rules through membership of ICOM, the access to funds through ICOF, but largely through finance they have raised for themselves, membership of ICOM has risen dramatically in the past three years. In 1976 alone twenty-two industrial co-operatives were registered, which was more than the total of registrations for this type of co-operative for the previous thirty years. By 1980 ICOM had seventeen member firms, over 300 associate members (co-operatives registered under the model rules) and some 300 individual members.

Most of the co-operatives registered with ICOM are small and innovative in the trades they pursue: for example, computer software, wholefoods, bakeries, craft workshops, self-build housing, bicycle hiring and electronics. They have been started mainly by young people and are more concerned with changing lifestyles and with the explicit social function of enterprises than were the original ICOM membership. Their presence within ICOM represents a radical and dynamic new force which the founders of the movement could scarcely have anticipated.

ICOM, which is still a very small organization with only a handful of full-time workers, has been developing regional groups through a full-time field officer, Bernadette Masterson. Six are currently operating – in London, Oxford, West Glamorgan, West Midlands, East Midlands and the north of England – and four others are in various stages of being set up. One of the most recently formed groups, ICOM North, is already well established and employs two full-time workers – David Connolly and Mike Holbrook. They serve a wide area from Newcastle across to Cumbria and, like the other regional groups, are building up local support for co-operatives and common ownership principles and establishing links with local trades unions, the Workers' Educational Association (W.E.A.), local authorities and local pressure groups. Much of their time is spent providing advice and assistance to local co-operatives and individuals wishing to set up businesses under the model rules.

Beechwood College, near Leeds, is now the headquarters of ICOM and a residential study and research centre for people working, or planning to work, in co-operatives, trades unions and local authorities.

In view of ICOM's success in fostering new enterprise – its membership is growing at the rate of two a week – it is deplorable that ICOM's small government grant ceases in 1981. At a time of growing unemployment and industrial and commercial decline, ICOM is a valuable national resource which deserves continued support.

There are signs of a growing relationship between the established (consumer) co-operative movement and the rapidly multiplying new breed of workers' co-operatives under ICOM. In July 1978 the Co-operative Bank launched a scheme to provide half the starting capital needed by workers' co-operatives. The Co-operative Union has become particularly active in the north of England, for example by helping to set up local and regional advisory services for small producer co-operatives. It recently helped three workers' co-operatives to get started in Skelmersdale, Lancashire, following redundancies caused by the closure of the Courtaulds and Thorn Electrical Industries factories. Under

Co-operative Union guidance, an umbrella holding company is being formed to channel some of the profits from the three workers' co-operatives to help existing co-operatives and start new ones.

Since 1976 the growth of small organizations established by local communities exclusively to promote and advise local co-operatives has blossomed throughout the U.K. There are at least seven firmly in existence: Devon and Cornwall Co-operative Development Agency, East Midlands Association of Common Ownerships and Co-operatives, Lambeth Co-operative Development Agency, Mid-Glamorgan Co-operative Development Agency, Northern Regions Co-operative Development Agency, Scottish Co-operative Development Committee, and Wandsworth Housing Development Agency; with a further forty or so preparing to set up. Collectively they have become known as local Co-operative Development Agencies (C.D.A.s).

They have been started, quite spontaneously, by people within local communities – members of local co-operatives, representatives of co-operative retail/wholesale societies, trades councils, local authorities, community workers, local branches of the Council for Voluntary Service, and individuals interested in co-operatives – who recognize that, if local co-operatives are to be encouraged and are to prosper, they will need specific local advice and expertise to meet their needs. There has also been a strong desire to set up democratically controlled agencies, organized and administered (and, it is hoped, eventually financed) by local co-operatives.

The Scottish Co-operative Development Committee (S.C.D.C.), for example, was formed largely through the efforts of the Scottish Council for Social Service (S.C.S.S.) in 1976, following a public meeting held to discuss the idea. Its objective is to encourage the growth of workers' co-operatives in Scotland through advice and training schemes in conjunction with the W.E.A., and publicity and liaison activity with local authorities and the Scottish Development Agency. The S.C.D.C. employs a full-time worker, Cairns Campbell, on an annual grant under the I.C.O. Act. A 'register of talents' is being compiled: a register of like-minded people with complementary skills, interested in coming together to form co-operatives. By 1978 there were only five workers' co-operatives in Scotland; in January 1979 there were fifteen (most of which had registered under the ICOM Model Rules). This increase is due in no small measure to the efforts of Cairns Campbell and the S.C.D.C.[61]

The S.C.D.C. has the longest track record of any of the C.D.A.s (most of them have started up in the past two years) but it is exceptional in the size of area which it covers. The majority of local C.D.A.s are indeed local. They purposely limit themselves to one or more contiguous local authority areas. By no means all the local C.D.A.s restrict their activities to promotional and lobbying efforts. They vary considerably in their aims, motivations and origins. A number of them are starting up revolving loan funds and the Wandsworth group was set up specifically to assist housing co-operatives.[62]

The funding of local C.D.A.s has come from a variety of sources. The Scottish Co-operative Development Committee receives funds under the I.C.O. Act, but these are very limited and it is unlikely there will be any more. The Northern Regions C.D.A. has had grants from the Tyne and Wear County Council (£5,000) and the Co-operative Union (£500). At least one local C.D.A. has received a grant under the Urban Aid programme and many others have made similar applications, usually to employ one or more full-time staff and, in one instance, to set up a loan fund.

In the future the main source of local C.D.A. funds is likely to be the local authorities. They have considerable powers to make grants to entrepreneurial organizations such as local C.D.A.s, for example, under Sections 137 and 111 of the Local Government Act 1972; and, under the Inner Urban Areas Act 1978, a designated inner city district council can grant a local C.D.A. up to £1,000 towards the expenses of helping each co-operative. Until such time as local C.D.A.s can become self-supporting through the contributions of member co-operatives, they will have to rely largely on grants from central and local government, charities, trades unions and retail co-operative societies.

Outside the more formal structures of local C.D.A.s, small co-operatives and co-operatively run groups (which are not legally registered as co-operatives) have also begun to organize their own self-help, education and training projects.[63] Many of these are informal gatherings of local and regional co-operatives, while others have a more formal and, in some cases, specialized, purpose for co-operatives in building, printing, wholefoods and so on. Skill exchange schemes between groups of co-operatives and communes are now widespread. SENSE (Skills Exchange for a Stable Economy), for example, is one of the organizations run on a voluntary basis to publicize training courses available to individuals and co-operatives and has drawn up a list of contacts for a wide variety of skills – from plumbing to goat keeping. There are also numerous projects being run to put similar co-operatives in different parts of the country in touch with each other and make widely known the skill needs, products and aims of co-operatively run enterprises.

A more specialized educational, information and resource service is offered by Workaid, an independent group of business and financial experts which provides short courses and advice to co-operatives in financial and organizational management. Similarly, Commonwork has been very active in sponsoring education on co-operatives and in bringing together individuals and organizations which aim to provide regional and national support for co-operatives (e.g., the Co-operative Development Agency and the Open University Co-operative Research Unit).

Small community groups are taking steps to support co-operative enterprise. For example, the Neighbourhood Council in Lewisham, London was instrumental in setting up a co-operative in the clothing trade, and in Brent, London, the local tenants' and residents' association has been actively pursuing plans to set up a local C.D.A.

There are other organizations that are not affiliated to the 'small-scale co-operative' movement or ICOM or the Co-operative Union which are also seeking to help co-operative projects. The Action Resource Centre (A.R.C.) places management 'secondees' from large companies with community projects that are involved in creating long-term work opportunities and need professional help and business skills. A toy-making workers' co-operative, House of Lambeth, established by the Lady Margaret Hall Settlement in London, has been helped in this way by a secondee from Rank Xerox. A.R.C., which is a registered charity, has been operating since 1973 and is building up a regional structure and major community schemes to cover secondees in all parts of the country.

Inwork, a non-profit company started in Fife in mid-1979 by

John Morrison of the Fife Regional Council, aims to start up and support small businesses in Fife. It will take a majority shareholding in some of the small businesses it establishes; others will be set up as charitable trusts to give work to a fifty-fifty mix of able-bodied and the handicapped; others will be set up as workers' co-operatives. For its support and advisory services on management, finance, marketing and so on, Inwork will tap the resources of industry, academic institutions and the Scottish Development Agency.

The Mutual Aid Centre (MAC) was set up in 1977 to encourage and promote small-scale co-operatives amongst consumers.[64] It aims to set up demonstration projects showing how the consumer movement can be extended to work on the local scale. MAC has already helped start a parent-teacher co-operative primary school in Maddinglay, Cambridgeshire; a co-operative repair workshop in Shrewsbury; and a mutual aid services centre run by the tenants of a tower block in Birmingham. A motorists' co-operative was started in early 1979 and a co-operatively run garage and workshop was opened in Milton Keynes in August of that year. Another MAC project, Brass Tacks, was started in April 1980. This is a furniture and electrical goods refurbishing and recycling community workshop in East London which employs twenty young people.

Based at the Open University is the Co-operative Research Unit (C.R.U.), which is investigating the political and management problems of setting up and running co-operatives. Techniques of accounting, costing and ways of testing the viability of co-operatives are being developed. C.R.U. also undertakes 'action-research' for ailing co-operatives and is compiling a library and computerized index of specialist publications and news cuttings on co-operatives.

Job Ownership Ltd (JOL) was formed in November 1978, under the chairmanship of Jo Grimond M.P., to promote 'the ownership and control of industrial and service enterprises by the people who work in them, rather than by the people who supply them with money'.[65] JOL offers a package of services to assist the formation of workers' co-operatives and the conversion of private companies to co-operatives, and will advise on ways to structure these enterprises, drawing principally on the Mondragon model of co-operatives in northern Spain.[66]

JOL has secured a two-year grant from the Rowntree Social Service Trust, but aims to become a self-supporting, non-profit making company through fees charged for its services. It is a non-political organization which its Director, Robert Oakeshott, sees as complementing the work of the national C.D.A. and similar advisory groups serving co-operatives.

The Institute for Workers' Control (I.W.C.) is a less specialized body which advocates and supports all forms of co-operative ownership and worker control in industry. Although the I.W.C. does not have sufficient resources to give detailed advice to workers' co-operatives, it has helped them with general advice and publicity. The I.W.C. has taken a lively interest in the *Scottish Daily News*, K.M.E. and Meriden co-operatives and in the principles of socially useful production and the Lucas Corporate Plan. It has published books and pamphlets on these topics and on all other aspects of industrial democracy and trades union strategies in the U.K. and overseas. I.W.C. also publishes the bi-monthly journal, *Workers' Control*.

Local C.D.A.s were given a major fillip in 1978 when the national Co-operative Development Agency (C.D.A.) was created by an Act of Parliament with all-party support. The Act made available £900,000 over three years (with a possible further £600,000) to the C.D.A. to encourage, advise and represent the interests of co-operative organizations of all kinds.[67] It also acts as a forum for discussion and debate within the co-operatives movement and advises government and public authorities on matters pertaining to the needs of co-operative enterprise. Though the C.D.A. acts only in promotional and lobbying roles (it cannot dispense capital grants to co-operatives), it has extensive advisory, research and training facilities – the latter based at the Co-operative College, Loughborough. The C.D.A. under its Chairman, Lord Oram, and Director, Dennis Lawrence, is especially anxious to develop co-operative organization in the fields of industry, housing and credit supply. It is also helping agricultural and fishing co-operatives to expand their range of activities and is exploring the possibilities of transport co-operatives and co-operatives for students and the disabled. To this end it undertakes feasibility studies on co-operative projects on a fee basis.

The national C.D.A. is as new as many of the local C.D.A.s, and since its formation it has been a keen supporter of moves by local groups to set up local C.D.A.s at county and district level. However it remains to be seen how and in what ways the national agency will respond to a strongly democratic local C.D.A. movement which is, quite literally, growing up under its feet. Similarly, the relationship between the local C.D.A.s and the regional ICOM groups has not yet developed to a point where it can be readily defined. The same must be said of the filiation between the established co-operative movement (and the national C.D.A.) and ICOM. Even so, for all these organizations, the area of common interest is very wide and the points of contention and possible conflict few. They all face a movement within their own ranks that is mostly composed of young people who are radical in their outlook and assertively democratic. The direction and reorganization of ICOM and the national C.D.A., for example, will be largely determined by this young and ebullient new force.

Although certain government agencies are beginning to display a more benevolent disposition towards worker co-operatives, ideological prejudice against them remains strong. The Manpower Services Commission has set up a Co-operatives Unit and is closely associated with the National Workshops Association, which was set up in 1978 to provide a nationwide advice, support and information service to small workshops, especially those set up under M.S.C. schemes. But the Manpower Services bureaucracy is far from enthusiastic about setting up co-operatives under the Special Programmes.[68] However, in the past they have helped to fund a few small and very successful co-operatives – for example, Dowlais Knitwear in south Wales and Wandronics in Wandsworth, London. The regional Development Agencies[69] and the Council for Small Industries in Rural Areas (COSIRA) have all expressed a willingness to help co-operatives get started but, as yet, they are not prepared to recognize the unusual circumstances of co-operatives and insist that they be treated on a par with other small firms. Neither have any of these agencies done much in the way of stimulating co-operative enterprise.

The same cannot be said of the Highlands and Islands Development Board (H.I.D.B.) in Scotland. It has aided workers' co-operatives in specific industries for several years and, in 1977, launched a scheme to encourage community co-operatives, or Co-Chomuinn, in the Western Isles.[70] Two field officers, Coinneach MacLean and Angus MacKay, have been stationed in Uist and

Barra, and in Lewis and Harris, to advise communities on how to establish these multi-purpose co-operatives. A handbook, in English and Gaelic versions, setting out the guidelines for establishing a Co-Chomuinn has been produced and distributed by the H.I.D.B. throughout the Western Isles. The H.I.D.B. will make grants towards the setting up of a Co-Chomuinn, matching pound for pound the capital raised by the community, and will pay for a full-time manager for three years.

By March 1979 four Co-Chomuinn had been formed; in Ness, Park, Yatersay and Eriskay. They have plans for a wide range of activities (some of which are already under way) in agriculture, fisheries, knitwear, crafts, retail trading and tourism. Co-Chomuinn Nis Ltd (Ness Community Co-operative), which covers 2,500 people in the fifteen townships of Ness (who have raised £16,000 for the co-operative), appointed a manager, James MacLeod, in 1978. He is working to set up a market garden – originally started as a horticultural nursery by the very active Ness Community Association under the Job Creation Programme – and assess the feasibility of other projects; a garage, workshop and bakery. Co-Chomuinn Nis has taken over some of the projects of the Community Association: the hiring out of modern farm machinery and the bulk purchase of fertilizers for crofters. Other projects include a co-operative store for agricultural supplies, and a building team.

The scheme for Co-Chomuinn has been well received in the Highlands and Islands and at least four more communities are likely to form co-operatives in the near future; twelve others in the Western Isles and Shetland and Orkney Islands have expressed a firm interest in the idea.

If the scheme proves a success in the Western Isles, it will be extended to cover the entire area served by the H.I.D.B. Much depends upon the enthusiasm and energy of local communities in identifying their problems and seeking remedies to them through self-help and community organization.

Community Initiatives

One of the most impressive examples of community self-help is the Craigmillar Festival Society. From the 1930s, Craigmillar – a large council estate housing 25,000 people on the outskirts of Edinburgh – has suffered from a high incidence of poverty, unemployment, housing neglect and totally inadequate social and educational amenities. In 1962 a local mother, Helen Crummy, dissatisfied by the lack of educational opportunities for her young son, suggested to the local school that an annual local festival could be organized to provide a display of the creative and artistic talent of the children and local people in the area and help to combat the unfairly critical press coverage given to Craigmillar. The idea won approval from other local residents and the Craigmillar Festival Society (C.F.S.) was formed.

The annual festival was organized and run voluntarily by them, with the local school headmaster acting as adviser, for seven years. In that time, the festival became more than a vivid and colourful demonstration of the artistic talents of the area: it developed as the focus for social change within Craigmillar as local people increasingly found their intentions thwarted by the lack of facilities for the festival – no meeting rooms, rehearsal rooms or exhibition halls existed in the area. They began campaigning for improved amenities in housing, employment and education. 'The Festival proved to be the key that unlocked the creative talents within the community and provided the field force for social action.'[71] The reaction of residents in Craigmillar, all of whom are automatically members of the C.F.S., brought them into a close relationship with the local authority, M.P.s and a wide range of outside institutions as they began suggesting ideas and setting up voluntary projects for community welfare.

In 1970 the C.F.S. obtained funds under the Urban Aid Programme for a Neighbourhood Workers' Scheme. This gave them resources to employ local people and spend £1,000 on a community project fund. In 1972 the local authority financed Pilot Scheme projects in housing, environmental improvements, social welfare, education, employment and arts. Under the Pilot Scheme, the area acquired a community centre, library and a high school – facilities which it had lacked for over fifty years. The C.F.S. also sponsored a very large number of Job Creation Projects.

The C.F.S. applied for funds to the E.E.C. Programme of Pilot Schemes and Studies to Combat Poverty, which was launched in 1975. Their declared objective was to work together with a team of local people and professionals, officials and politicians towards the further improvement of conditions on the Craigmillar estate and, at the same time, to study any major obstacles to their plan. They became the only self-organized community group in Western Europe to get an award under the Programme. The funds have been applied to projects in housing, employment, social welfare, education and transport. These are seen as innovative experiments in local responsibility which, if effective, will create their own basis for continuation and expansion. They have proved so successful that the E.E.C. grant was renewed until 1980.

Since 1962 the 'midget' of the first Craigmillar Festival has become the 'gentle giant' – symbolized in an enormous children's play structure built by the C.F.S – of the present Festival Society. Over 150 full-time, part-time, and voluntary workers run or help to organize a quite breathtaking range of community projects: youth clubs, social programmes for the elderly and handicapped, arts and drama activities, play-schemes, industrial and neighbour-hood workshops, community transport, professional and student secondments, a community lawyer and a community printing service.[72]

In 1978, with help from the Scottish Action Resource Centre, the C.F.S. started its own company, Craigmillar Festival Enter-prises Ltd, which began trading in January 1979 from a site donated by Esso Ltd. It employs eight former long-term un-employed people in building trades, maintenance and construction contracting, organized and run by George Burt, the manager. The company, a registered charity, hopes to expand its workforce to eighteen as more work becomes available from local authority contracts.

A monthly newsletter of the C.F.S. is published – and circu-lated to a wide audience outside and within the area – giving up-to-date details of the Pilot Scheme projects. The bi-monthly *Craigmillar Festival Newspaper* is delivered free to all houses in the area.

The annual festival continues to be held in Craigmillar, starting in midsummer week and continuing for two to three weeks. It is still an integral part of the Society which draws much strength and cohesion from its colourful exuberance and it serves as a constant reminder of the vital role arts and music have played in refreshing community spirit within the area.

In November 1978, the C.F.S. published *Craigmillar's Comprehensive Plan for Action*.[73] This draws together analyses and recommendations for shared government and partnership between the community and central and local government and outside institutions in tackling local objectives in each of the areas covered by Pilot Scheme projects. The plan is a vision for the future, 'the achievement of a viable community with all the necessary ingredients of amenities, facilities and services . . . Its basic importance lies in the fact that it advocates and signifies a change in politics and economics to yield a more fulfilling society.'[74] The District and Regional Councils have agreed to consider and implement the plan in conjunction with the C.F.S.

Copies of the plan have been distributed to China, the U.S.A., Russia, Israel, Canada and India and throughout Europe – the imagination, vivacity and local creativity that has blossomed from a community arts festival in 1962 is seen as an example with worldwide significance.

In the C.F.S. local people identified their own problems and needs and set about tackling them. Their campaign began on a very small scale, with the single rallying point of an arts festival, from which it steadily expanded into other fields of social action.

The Craigmillar Festival is the leading community initiative in Britain, but during the past five years many other communities throughout the country have decided to take a hand in shaping their own economic future.

An example from the north of England is the Askrigg Foundation in Cumberland. Askrigg is a village of 400 people in the Pennine hills. A few years ago the vicar, Malcolm Stonestreet, formed the Askrigg Foundation to be a launching pad for a series of local initiatives: a youth centre that runs residential courses and is used by some 2,500 young people a year; a village shop and coffee house, run by Askrigg Industries; an old people's home operated by the villagers' housing association; and Askrigg Arts, another community group, to promote cultural activities. Negotiations with the Development Commission have secured a mini-factory for the village, to accommodate local craftsmen and small manufacturers. In all these ventures the Foundation takes the first steps, and then passes control and when possible ownership to the community. They are now exploring the possible applications of mini-hydro electric plants in the locality.

Among several community initiatives now starting up in Wales, there is the Cymdeittas de Gwynedd, a co-operative near Bala in north Wales. As elsewhere in rural Wales there is rapid migration of young people, disappearance of small farms, very few new job opportunities. The co-operative grew out of a series of public meetings held in the mid-1970s about the problems of workless-ness in the area. It is run by a committee of twelve local people under the leadership of Wynn Edwards, an experienced business-man and co-operator who manages the enterprise. They now run a cash and carry and freezer store, and three refrigerated vans serving some 400 shops in the region. All the capital was sub-scribed locally – there are 250 subscribing members – and in 1978 the co-operative had a turnover of three quarters of a million pounds.

This venture could well be a model for other parts of rural Wales. Their current plans include building a greenhouse for bedding plants, and starting a fruit processing unit. They envisage several other activities, such as installing a local sawmill, and the manufacture of wooden building components from local timber. Ultimately they hope to purchase land and housing, to regenerate small farming and horticulture, for which the co-operative would provide processing and marketing facilities.

The initiatives in Craigmillar, Askrigg and Bala are three of about thirty community efforts on similar lines, that is, where there is a broadly based community involvement aimed at raising the real income of its members in a variety of ways. During 1980 the Foundation for Alternatives produced a two-volume set of profiles of community initiatives, including the three referred to above, and the Foundation is now promoting the interests of community initiatives and forming links between them.[75]

There are at least another thirty local initiatives that are more specifically concerned with promoting industry, business and employment in their communities. These are Local Enterprise Trusts (LETS). They are varied in their constitution and in their methods of operation and only a few of them are trusts in the strict legal sense. They do, however, carry certain hallmarks which distinguish them as a group and which are included in their definition.

First, they are broadly based community organizations. This

distinguishes them from a service provided by a firm or by the government. They are associations which mobilize the enthusiasm and goodwill of local people and institutions (including local industry) to encourage small-scale wealth and work-creating activity. As an organ of the local community, LETS are able to attract and channel this goodwill.

Second, they are necessarily local, though the size of the locality could vary from a village and its neighbourhood to a sizable town or rural district. Their work is based on personal contact and intimate first-hand knowledge of the local area and its inhabitants.

Third, they are themselves enterprising organizations. They are not counselling bodies which wait for approaches, though they do certainly have a counselling role. In the true tradition of British voluntary organizations they go out and look for trouble and take initiatives themselves to discover and respond to the problems affecting small enterprises in the area.[76]

Some LETS have local authority support and others are starting with help from local industry, and with loans and grants from the Urban Aid Programme and from the Manpower Services Commission. Few of the LETS are more than two or three years old and, in many cases, their structure and aims are innovative and experimental.

Since 1977, John Davis of the I.T.D.G.'s Appropriate Technology for the U.K. project, has been engaged in identifying such groups, helping their formation and forging links between them. In association with the Foundation for Alternatives he has started a three-year project to link together six LETS in different environments and of different backgrounds to channel resources to them where they are in need and to mobilize a body of knowledge based on their practical experience over the period. Besides providing a guide to action by all other community groups, this study will make possible an assessment of the services available to small enterprises, whether government or non-government. There follows a brief outline of the six LETS selected for special study under the I.T.D.G./Foundation for Alternatives project:

Antur Aelhaearn (Wales)

Llanaelhaearn is a rural community of some 300 people in the Lleyn peninsula of Wales. In 1971, led by their local doctor, Carl Clowes, they started to reverse the decline of their community. Four years later they formed the first community co-operative in Britain. The Antur is based on the principles of self-help, co-operation and a culturally and economically viable community. They have built their own work-centre, which now houses twelve full-time employees working on knitwear. Forty more people work part-time in their homes. The knitwear factory, established by the Antur, reached a turnover of £50,000 in 1979.

For the future, they are planning a rural industries centre to provide information on the Antur to the increasing number of callers and potential entrepreneurs and advice on how to set up similar co-operative ventures; and to display an audio-visual exhibition of the history and industrial background of the area. They are also working on a community centre.

Clyde Workshops Ltd (Glasgow)

Clyde Workshops Ltd was set up in January 1979 as a subsidiary company of the British Steel Corporation to provide and manage workshops and offices let at commercial rents to small firms. A back-up management advisory service is provided free to the tenants. It was set up by the B.S.C. to create viable employment in an area suffering from high unemployment as a result of steel plant closures. After a few years, the company intends to hive it off as an independent venture. It already operates without subsidies.

The response to the scheme has been rapid and very enthusiastic. Within nine months of opening the available accommodation was fully let to sixty-five companies (fifty-three of which were first-time enterprises), employing over 500 people.

The chief executive of Clyde Workshops, Stewart Anderson, and the B.S.C. are also considering developing a separate and parallel company offering a consultancy service for the rehabilitation and conversion of buildings for workshops, small factories and retailing units; product viability studies; and design and layout advice for small firms.

Hackney Business Promotion Centre Ltd (London)

This was set up in 1979 in East London as a voluntary, non-profit making organization with Dennis Statham as Manager, jointly supported (but not controlled) by local industry and local and central government. It aims specifically to assist small businesses, of whatever structure, and encourage others to start. One of its first ventures was to launch a programme of one-day conferences, 'Start Your Own Business', with intensive follow-up courses.

Since its early days when it was primarily a counselling service, the Centre has widened its range of activities to include more training programmes, the conversion of a warehouse into small workspaces, activity among the ethnic groups of the area, an advertising campaign to attract new businesses, the organizing of a business to business exhibition and the launching of industry associations. The Centre also acts for the local council as financial adviser in its dealing with local firms. A series of ten educational programmes for small businessmen screened in autumn 1979 was partly filmed at the Centre and featured some of the young entrepreneurs taking the Centre's business courses.

Business Link Ltd (Runcorn)

An organization formed on the initiative of the Business Opportunities Project run by I.C.I. Ltd and assisted by local government, development agencies and banks, Business Link aims to encourage the growth of new business and employment in the area and act as a focus for community enterprise. As an enabling organization for local enterprise, the intention is to provide expert help to small businesses through staff seconded from large companies in the area. In the first six months of its existence, Business Link handled over 130 enquiries from local firms.

Community of St Helens Trust (Merseyside)

The Trust was set up in 1978 as an independent local body by members of the community, trades unions and local government. The major employer in St Helens, Pilkington Bros Ltd, gave substantial support to the Trust, as did a number of other local companies, and the Town Hall and local authority have been extremely helpful.

The Trust aims to create employment in the St Helens area by assisting new or expanding businesses. This is done directly or through the Trust's supporters, giving financial and professional advice (and very occasionally small loans), premises, business opportunities, training and generally bringing people and organizations together. By mid-1980 the Trust had been involved in the creation of 800 full-time and 200 part-time jobs. The Trust's Director is Bill Humphrey.

Somerset Small Industries Group

This was formed in 1977 by four local residents as an unincorporated association, whose aim is to promote forms of small industry appropriate to a rural or village setting in west Somerset. A full-time employment officer, Fred Wedlake, was appointed, paid for by the M.S.C. Other financial support has come from I.B.M. Ltd, B.P. Trading Ltd, Shell (U.K.) Ltd, the Midland Bank and the Foundation for Alternatives.

Since their formation this group has helped over thirty small firms and individuals to find premises, boost sales and generally extend their range of business contracts. They have also held several public meetings and seminars in towns in west Somerset. An annual local trades fair is run by the group and they have recently produced a comprehensive local trades directory.

Working communities are an extension of the principle which motivates the organizers of LETS and, since 1970, they have begun to spring up in many inner city districts. They are groups of independent small firms co-operating in sharing a building and joint services. The buildings they occupy have been converted to small unit tenancies – offices, workshops, design studios and so on – from disused and unwanted large factory sites or warehouses.[77] Characteristically the working communities have set out to select as members small firms (many of them one- and two-man new enterprises) which have something in common, so that skills or products can be linked and used within the building to mutual advantage.

There are seven working communities in London alone that have created workspace for over 900 men and women; Clerkenwell Workshops, the largest, provides small work units for 350 people in a converted School Board bookstore. The London-based communities have set up the Federation for Working Communities to help establish similar projects throughout the U.K. and popularize the ideas behind them.

5

A GUIDED TOUR
OF ALTERNATIVE
ORGANIZATIONS IN
THE U.S.A.

It is no accident that it is in the world's richest country, where monster technologies and their supporting institutions are most dominant, that there also exists a most flourishing alternatives movement.

It is here that the disfiguring impact of large-scale, capital and energy-intensive industrialism is most evident; and the vulnerability of the whole structure is daily becoming more obvious. But the other side of this threatening picture is the emergence of a movement whose starting point is individual and community self-reliance, and whose aim is to introduce non-violent and sustainable technologies and life-support systems.

The spontaneous flowering of this movement will doubtless be a matter of interest to future social historians. It probably owes something to the fact that the American education system exposes a higher proportion of young people to higher education and the world of ideas than is the case in most other countries. Again, the 'cultural memory' of most Americans has probably something to do with it: few of the families that I have met in the U.S.A. have to go back more than one or two generations to recall a time when they were far more self-reliant than they are now; whereas in the Old World this folk memory is dim, if it exists at all. For these reasons alone the warning signals of people such as Wendell Berry, Murray Bookchin, Kenneth Boulding, Rachel Carson, Barry Commoner, Hazel Henderson, Ralph Nader, Barbara Ward and many others were clearly received and taken to heart. It is certain, too, that Schumacher's *Small is Beautiful* gave substance and conviction to the widespread intuitive feeling that we are heading in the wrong direction, and encouraged many people to strike out on their own towards new lifestyles and a saner society.

Whatever the reasons, the fact remains that within the space of a few years the alternatives movement in the U.S.A. has already reached such dimensions, both qualitative and quantitative, that it almost defies description. One directory published recently, *Alternative America*,[1] lists over 5,000 organizations; another, which deals more specifically with appropriate technologies – the National Science Foundation's *Appropriate Technology: A Directory of Activities and Projects* – suggests that at least 300 groups of innovators are working on technologies that reflect a concern for the preservation of ecological balance, the recycling or cutting down of waste, and the fostering of local self-reliance.

In this chapter we have not tried to be as comprehensive as we have been for Britain – the sheer number and variety of the U.S. groups makes this virtually impossible – but rather to describe what is being done by the leading organizations in the field, and at the same time to convey to the reader something of the quality of what is being done and who is doing it.

With this is mind, towards the end of 1978, we made a tour of alternative technology groups in the U.S.A. on a selective basis. This means that we have missed out many important groups, but in terms of the range and scope of what is being done, we believe that our survey is reasonably representative. Our route took us from Maine to New York, via Boston, San Francisco and Washington.

Our tour started at Forest Farm, on the coast of Maine, near Harborside, the home of Scott and Helen Nearing. For nearly fifty years they have lived off the land in Vermont and Maine, in homesteads which they have built with their own hands. Helen is seventy-five and Scott ninety-five. They combine practical self-sufficiency with a vigorous intellectual life: they have written some fifty books on economics and politics, homesteading and organic gardening, and are in great demand as lecturers. Their farm has become an information centre where interested homesteaders, organic farmers and others can seek practical advice, much of which is now in the form of low-cost publications.

The Nearings are virtually self-sufficient in food, no mean achievement in the hard Maine climate. Scott grows all the vegetables they need, and more, on a half-acre garden of incredible productivity. The secret of its fertility, he says, is the tons of compost he makes every year, using seaweed, and soil accumulated in a shallow pond. The half-acre garden includes a large greenhouse into which Scott transplants vegetables in the autumn, for winter use.

As a socialist and a pacifist – views that forced him into leaving academia and seeking self-sufficiency fifty years ago – Scott is concerned with the survival of a society he sees under the threat of large-scale capitalism. These dangers, and the alternatives open to individuals and communities, are the subject of his book *Man's Search for the Good Life*.[2] In their best known work, *Living the Good Life*,[3] the Nearings describe their own work in building, organic gardening and vegetarian living. 'Life's necessaries', they say, 'are easily come by if people are willing to adjust their consumption to the quantity and variety of their products.' Scott and Helen Nearing are two great teachers who have developed high intellectual and manual skills by living in total harmony with nature. As we left their hospitable farmhouse, Scott resumed the work we had interrupted, digging the foundations of another greenhouse.

Riding down to Bangor, our next stop was at HOME (Home Workers Organized for More Employment), a resource centre for rural community development. It was started by local people in an effort to improve their standards of life through local self-sustaining development. It runs four development programmes: a community learning centre; a crafts marketing co-operative; a farm marketing co-operative; and a land trust, Self-Help Family Farms.

One of the objects of the learning centre is the teaching of craft skills as a means of supplementing local incomes. They also run classes for high school diploma credits. Of the 400 students who attend the centre and who range in age from sixteen to sixty-five, eighty are working for diplomas.

The marketing co-operatives comprise a group of small shops which provides an outlet for locally produced crafts and farm produce. The creation of these marketing outlets has led to a revival of cottage industries and a strengthening of family self-support.

The land trust enables local people to build houses, grow food and raise livestock in an environment in which land prices and construction costs would otherwise be prohibitively high. By leasing land from the trust, a poor family can get a farm on the understanding that they will contribute their labour to the building of other people's homes. In this way energy-efficient houses have been built costing $10,000 or less. Low monthly payments cover mortgage, taxes, insurance and land rent. The trust is now exploring the possibilities of community investment: thus, say, $100,000 donated to HOME could be repaid in ten years by small farmers who could be settled with the aid of this money and grow timber as a cash crop for purposes of repayment.

Altogether some 2,000 people use HOME's services throughout the year.

On our way to Bath, Maine, we enjoyed a working dinner with Bill and Katie Thompson, a young couple who had recently left Boston and settled in a Maine village, willingly taking a large cut in income in order to live more simply but also, as they put it, to become part of a real community. They are part of the growing movement towards voluntary simplicity in the U.S.A. It is estimated that some 3 million people have done something on similar lines, and that by the end of the century there could be upwards of 50 million doing the same, deliberately going for a low-consumption, more socially co-operative, life-style. (Critics may say that the people who practise voluntary simplicity can well afford to do so. But then, they could well afford to abstain from doing so, too, but haven't.)

On our way to the Thompsons' we called in on a like-minded friend, Bill Myers, a marine biologist who is determined to upgrade ways of getting a living from the sea. He runs Abandoned Farm, a small mussel farm on which, within a couple of years, and using very simple equipment combined with a lot of biological knowledge, he has greatly improved both the quantity and quality of mussels that can be farmed on a very small area: a skill and knowledge-intensive method of making a small (three to four people) operation possible.

At Bath, Maine, there is the Shelter Institute, a resource or service centre set up four years ago to provide people with all the knowledge, both practical and theoretical, that they need to build their own homes. Their aim is to offer alternative methods of dealing with all the main building problems, and the long- and short-term implications of the alternatives. Thus all common methods of farming, wiring and plumbing are taught, along with other methods which free the student from standardized and expensive commercial practices. These basic courses total sixty hours and combine classroom work with practical work on local houses. They teach some 800 people a year and students come from all parts of the U.S.; so far Shelter students have built more than 400 houses for themselves.

Shelter also runs a design seminar for students planning to build their own houses, and a cabinetry course on interior finishing. Besides a good library and bookshop, they also sell most of the hardware needed in a house.

Ten miles away, an off-shoot of Shelter, Cornerstones, is approaching the housing problem from a different angle. While Shelter concentrates on equipping the individual home-builder with all the relevant knowledge needed to make a really efficient job of it, Cornerstones is concentrating on the small building contractor. By no means everyone can or wants to build for himself; so, they argue, small contractors should be taught how to build as economically and energy-efficiently as possible.

They run about twenty classes a year with forty contractors – who mostly come from neighbouring states – in each class. They also run evening classes for local builders.

According to the *Maine Times*, Shelter and Cornerstones are unique in the U.S.A. in the services they are providing. (So indeed is the *Maine Times*, a democratically run weekly newspaper which is read widely outside the State of Maine for its detailed reporting and its excellent coverage of appropriate technology developments and applications.)

Moving to Massachusetts, our next stop was at Cape Cod, to visit what is deservedly one of the best known appropriate technology centres in North America: the New Alchemy Institute. New Alchemy was started by a remarkable scientist/doctor, John Todd, and his wife Nancy, in the early 1970s. Their research and development work now embraces aquaculture and intensive organic agriculture, solar energy, windmills and bio-shelters. Their aim is to design and test human support systems – food, energy, shelter – that are environmentally sound, low-cost, and capable of widespread use on a decentralized basis, appropriate for families and small groups. Their method is essentially to substitute knowledge of natural systems and biological processes for fossil energy and man-made capital. In these respects it is one of the most important pieces of development work in progress anywhere.

Visually, the most striking of New Alchemy's experiments are the bio-shelters. These are structures designed for northern climates, which combine solar heating, wind power, fish farming, gardening. The Cape Cod Ark – a model for the larger and more complex Prince Edward Island Ark – is an integrated system, providing year-round agriculture and aquaculture in a passive solar structure, that is entirely self-sufficient in energy. New Alchemy have shown that, for instance, such an integrated system can produce the phenomenal fish yield of 90,000 lb. per acre annually: the kind of yield that is obtainable in Java is being easily reached in the rigorous Cape Cod climate.

When the solar algae tanks in which the fish are reared are placed inside a solar greenhouse, the tanks absorb solar energy so efficiently that they pay for themselves on their heating ability alone; and no other heat source is needed. (In conventional greenhouses, fuel accounts for about half the cost of flower and vegetable production.) The high yields of fish come virtually as a net bonus.

Inside the greenhouse, wastes from the solar algae ponds provide nutrients and warmed irrigation water for high-priced off-season vegetables and flowers. In return the agriculture contributes some weeds for the fish to eat. Meanwhile the moist warm air inside the solar structure is ideal for tree propagation, and tree seedlings find a ready market. The potentialities of this kind of low-cost, high-yield, sustainable system are only just starting to reveal themselves. We shall hear much more about New Alchemy during the next few years.

Other projects on the New Alchemy campus include intensive biological horticulture, which gives yields four or five times greater than the U.S. average, outdoors aquaculture and agricultural forestry. They also run workshops for all comers on Saturday mornings, May to October, on different aspects of their development work. Members receive a newsletter and an annual journal, both very informative.

Our next stop was in Cambridge, where the tradition of revolutionary ideas is kept going by the Institute for Community Economics. For several years now this has been promoting a land trust movement, and they have recently launched a Community

Investment Fund. Both are aimed at giving more people access to the ownership and control of productive resources: this is the kind of institutional support without which appropriate technologies cannot flourish.

The Community Investment Fund is aimed at mobilizing the investment funds needed to democratize ownership and technology. It is set up primarily to finance democratically managed or co-operative enterprises, with most of its portfolio in fully secured or federally guaranteed investments. Unfortunately, by the early 1980s the Fund had secured only about half of the $3 million needed to get it going, and the project has had to be shelved for the time being.

The Institute's community land trust movement already has an established track record. In 1968 they started a land trust, New Communities, in Lee County, Georgia. Today New Communities operates a 6,000-acre co-operative farm near Smithville in Georgia, the largest black-owned single tract in America. Since then they have found a variety of ways of providing people with access to land. In Vermont, Earth Bridge Land Trust comprises fourteen leaseholds on five sites in four different towns. A land trust in Appalachia is being developed as a model of land conservation combined with small farmsteads. In Oregon there is a women's land trust providing access to traditionally landless groups: women, black sharecroppers and migrant farm workers. As Bob Swann of the Community Investment Fund says:

> Although political democracy has involved the majority in the decision making processes, this has not been true in the economic realm. The so-called 'self-correcting mechanism of the market' has consistently 'corrected' toward the benefit of the few rather than the many. The democratization of economic enterprises will bring about greater personal, social, environmental and political security wherever it is introduced.

Few would nowadays dispute the need for more decentralized ownership and control over productive resources, which is another word for local self-reliance. The Institute of Community Economics is showing how it can be done.

Still, in Cambridge, we looked in on the Centre of Community Economic Development, a support group for community development corporations, for which they do research and carry out field assistance; they are in touch with some forty of the 400 or so corporations throughout the United States. One of their current ideas is to set up industrial parks within cities, to house small businesses – a means of financing local services. In east Los Angeles, for example, they are working with a community union started by the auto workers and the community development corporation, with help from the unions and the Ford Foundation. They are working on housing development and a minority business investment corporation; they are also running education programmes on nutrition, and starting an industrial park with the aim of creating 3,000 jobs. The Goodyear Corporation is now providing some help, and the community union is planning a thrift and loan mutual savings bank. An aquarium manufacturing unit has been started as a joint venture with a private entrepreneur. Another example is the Central Country community development corporation in California, where a group of technicians are serving a number of community-based co-operatives, which include a strawberry co-operative and a greenhouse development.

Community development organizations of this kind have been growing rapidly in the U.S.A. during the past ten years. Many are owned and operated by minorities, and are introducing new forms of business structures and income-generating activities. Several times during our tour the names of highly successful community organizations cropped up: Mace/Delta in Greenville, Mississippi, the South East Alabama Self-help Association in Tuskeegee, Alabama, and in the south-west, the Guadelupe Organization and the Calexico Community Action Council. Our impression was that a growing number of community corporations and the like are recognizing that social welfare alone may keep people afloat but does not integrate them in society; so they are moving towards promoting new economic activity.

This point is well illustrated by the work of a community organization I visited about a year earlier, in Berea, Kentucky. Berea is best known for its college, which owns a number of enterprises in the neighbourhood, in which the students work part-time while they attend the college. It is also the home of the Mountain Association for Community Economic Development (MACED). It was formed about three years ago as an alliance of community based groups in Kentucky and Virginia. Their experience had led them to conclude first, that real change demands the building of *economic* institutions that serve the needs of people with low incomes – usually small, locally owned, job-providing co-operatives and local enterprises; and secondly, that starting small businesses requires not only capital but many skills which were often absent. So MACED was formed to provide these skills needed for business financing and development; and one of their principles is that they stay with their clients until these skills have been transferred to them.

Drawing on the services of 100 specialists, MACED have already provided technical assistance of this kind to more than forty local enterprises. In one case the result was a major expansion – doubling sales in four years – of Kentucky Hills Industries, a thirty-three year old wood products producers' co-operative. Other cases have included the starting up of a small furniture production shop, and a major expansion in the output and profits of the Bread and Chicken Shop, a seven year old worker-owned bakery/restaurant employing twenty-one women. The same kind of technical support is being given to local community organizations, for instance, in starting a house-building programme or setting up a community development credit union.

There is no substitute for providing people with the techniques of self-help. At our next port of call, Chicago, we were to see the same principle applied on a larger scale.

We started off at the Centre for Neighbourhood Technology (an offshoot of the Illinois Neighbourhood Development Corporation), where Stan Hallett, Jesse Averbach and Scott Bernstein took us through some of the many local initiatives in their city. They started with a brief diagnosis of the problems that required these new initiatives. In the days of 'limitless growth' Chicago's prosperity rested on three assumptions: creating more wealth was the problem; large-scale technology was the answer; large corporations the vehicle. A stagnating economy, and rising costs of production and environmental damage have negated these assumptions. Industry has been leaving the city, and each year thousands of houses disappear in the city and large numbers are being built in the suburbs.

Similar problems are hitting other cities. But it happens that Chicago was built on a swamp. The sewerage system needs completely renewing, and the cost of this would be prodigious: as an enterprise it compares with laying the Alaska pipeline.

Renewing the sewerage system on conventional lines would mean pouring vast sums of money underground, and thus saddling the community with additional costs but no new incomes. So the Centre for Neighbourhood Technology began working on a strategy that would lower costs and raise incomes.

They started by looking at the technology of the sewerage system. The existing technology assumes that storm rain and sewerage should be on the same system; and that it is a regional and not a local problem. The Centre are now tackling the problem of how to recover and utilize the waste – they calculate that the community is losing some $250 million worth of methane gas and fertilizer annually.

This approach has led to other proposals for improvements in the city, including the introduction of urban agriculture and greenhouses on vacant lots and rooftops, and a programme of building maintenance and conversion. A team which made a study of nutrition among the minority populations discovered that as local shops and cheap restaurants disappeared, nutritional standards declined. An obvious answer was to get urban food production going, an idea that has greatly appealed to many black people who came from the rural south.

One way of cutting costs significantly for the poor is by cutting their heating costs. To achieve this, an energy conservation loan guarantee fund was set up. Under this scheme, the banks lend money to the homeowner to encourage the conservation of fuel, chiefly by insulation, and a guarantee fund persuades the utility company to charge lower rates for the energy supplied to such a user. Because these home owners are spending less on their fuel bills, they can repay the loan over an arranged period.

The major income generating development was introduced on Chicago's South Shore: the South Shore Bank. This area had progessively deteriorated as businesses left and houses were run down or abandoned, until the point was reached when some 300,000 poor people, blacks and other minorities, were without any banking facility at all. So it was decided to buy a bank. This was done by the Illinois Neighbourhood Development Corporation, supported by charitable foundations, church groups and business men.

Today the South Shore Bank, with deposits of nearly $60 million, ranks in the top 12 per cent of commercial banks in the

U.S.A. Its development loans, which concentrate on housing and small businesses, amount to $15 million. They have fewer bad debts than most banks, and they are profitable; when we visited the bank it was packed with customers. With this remarkable operation the community have succeeded in linking finance, housing, new businesses and more work; and Chicago's South Shore is looking up again, and the people have much to be proud of.

Another offshoot of the Illinois Neighbourhood Development Corporation is the Neighbourhood Institute, set up specifically to work on the human side of local development. Their tasks are to identify the basic problems of neighbourhood development and ways of meeting them: who and where are the unemployed; what work is available, and possible; and what are appropriate forms of training. They have also started an energy store, selling solar energy equipment and services, and are beginning to train South Shore contractors to undertake the necessary retro-fitting for solar heating and greenhouse construction.

To see some of these ideas being put into practice, we visited the 18th Street Development Corporation, which works on housing renewal. It started with two single-family houses, which were refitted and sold; but thereafter it is keeping the buildings within the community, and the next two buildings are being renovated to be rented to larger families. (A subsidy in the form of free labour and the use of recycled materials is translated into lower than market rents.) The Corporation's chief function is to train young people in carpentry and they take about thirty young people at a time. The instructors are qualified journeymen who are paid union rates; there is one for every five trainees. When we visited them they had just completed one building, were starting on another, and had dug the foundations of a very large lean-to greenhouse on a third.

Our final visit in Chicago was to Operation Brotherhood, a community centre serving the needs of the elderly and physically handicapped. It is run by Belle Whaley, a local community organizer of great imagination and talent. Their motto is 'to provide, protect and teach', and they do all three joyfully and efficiently. They provide services for nearly 2,500 people on the West Side. They cook hot meals for 100 people a day, six days a week, at the centre, and deliver 300 meals a week to people in their homes. They provide employment for more than 100 elderly people as home-helpers, technicians, gardeners.

The centre also runs two food pantries which sell fresh food at cost price to the elderly: one at the centre and the other as a mobile grocery store which services over 1,000 people in the area. Many of the vegetables are grown on their own mini-farm, a one-acre plot inside the city limits, worked by seven elderly men. During our visit a solar energy team was constructing a rooftop greenhouse at the centre, to extend their growing season to the full year.

All the self-help and community groups in the Chicago region are very well served by a journal entitled *The Neighbourhood Works*, produced fortnightly by the Center for Neighbourhood Technology. Its starting point is that neighbourhoods are the right place for new 'low' technologies to meet human needs for food, clothing, shelter, work and a clean environment. Because many appropriate tools and techniques are already available and can be used by people in urban communities, *The Neighbourhood Works* provides detailed practical information on projects, ideas, and technologies drawn from over 100 publications. This digest of information and guide to practitioners is exactly what is needed by individuals and groups working towards greater self-reliance.

Our next stop was Minneapolis, where we began with a visit to the Control Data Corporation. Its president, William Norris, is one of the leaders of large companies who recognizes the wider social implications of the growing dominance of large firms, and Control Data are now taking a hand in building up the small family farm in the United States. Small family farms, using a minimum of fossil-based energy, Norris and his colleagues believe, can be shown to be more efficient, economically, socially and environmentally, than large-scale corporate farming. To prove it they are buying ten small farms in the neighbourhood, with the aim of identifying and introducing the minimum requirements to provide a family with a really good livelihood on a sustainable basis.

By involving itself in alternatives in this way Control Data is probably only a few years ahead of its time. A small but growing number of large firms on both sides of the Atlantic are looking very hard at structural unemployment, energy costs and other indicators that bode ill for conventional industrialization. It can only be a step in the right direction if large corporations begin to see themselves as trustees of national and local assets, and not simply as machines geared to maximizing profits for private appropriation.

From one of the biggest corporations in the country we went to one of the most successful co-operatives in the mid-West; two sharply contrasting ways of mobilizing human effort. The Distributing Alliance of North Country Co-operatives (DANCE) in Minneapolis has over 100 members, small co-operatives in the surrounding region, and an annual turnover of more than one million. DANCE's trading profit, most of which is returned to members, is about $30,000 a year.

DANCE started in 1970 when a small group of young people moved from a farm into the city and began buying their food in bulk to save money. More people soon gathered around, and they started a co-operative run from the basement of a church. By 1971 their turnover was about $50,000 a month. As more co-operatives grew up they started their central warehouse, and moved into non-food items as well. Most of the co-operatives are in the city, with shops for hardware, books, electronics, as well as bakeries and restaurants.

In the early 1970s they became incorporated and are now part of the established co-operative movement in the U.S.A. This marked a change from enthusiastic amateurism to enthusiastic professionalism.

The DANCE warehouse is also the headquarters of the All-Co-operative Assembly, which undertakes educational work relating to the co-operative movement, helps to start new co-operative ventures and provides technical assistance such as book-keeping and legal advice. This unit is supported by a small contribution from each of the member co-operatives.

Many of the co-operatives get a good deal of volunteer help, and DANCE have introduced a special volunteer membership, through which volunteers get concessionary prices from co-operatives depending upon how many hours of work they put in. One of the organizers of the DANCE warehouse is Jerry Chase, who works during the mornings with the co-operative and in the afternoons as a computer programmer (on Control Data equipment).

After a meal in a workers' co-operative restaurant, we drove out to Milaca, to visit Don and Abbie Marrier, who produce *Alternative Sources of Energy*. This was started as a newsletter in 1971 and is now a quarterly magazine, one of the best on alternative technologies. Its emphasis is on renewable energy sources such as solar, wind-generated power systems and small-scale hydro, and it is written by practitioners for practitioners. The Marriers also run a lending library as a service to their readers, specializing in practical and hard to find information.

The Marriers' home is heated by a combination of wood-stoves and solar energy, and all electricity is generated by a windmill. Don is working with Martin Jopp (one of the pioneers of wind power for electricity) to develop small wind generators to produce about 600 watts for household use.

The copper-mining city of Butte, Montana, came next on our itinerary. This is the headquarters of the National Centre for Appropriate Technology, brought into existence by the efforts of the former Senator Mike Mansfield. Its goals are to develop and make known help to introduce technologies appropriate to the needs and resources of low-income communities. Its creation and funding with Federal support (it is financed by the Community Services Administration) marks a growing recognition of the fact that the human and economic problems of the poor cannot be solved by welfare payments, but only by the opportunity to do work that makes them more self-reliant as individuals and more self-sufficient as communities.

The Centre, run by a team of young and enthusiastic staff, sees itself as a launching pad for small-scale, low-cost, self-help technologies that are within the reach of poor communities. It is building up a comprehensive information service on alternative energy systems and conservation, on intensive organic gardening, on co-operative and worker-ownership and other aspects of the hardware and software of appropriate technology. They also run a small research and development programme. This includes a solar energy house heating research unit which is run in collaboration with the Los Alamos Scientific Laboratory and is intended to produce specifications for architects and builders.

The Centre now has a staff of fifty. Much of their effort is devoted to sorting out and responding to the applications for grants under the Centre's programme for grant-aid for local appropriate technology development. The grants are mostly in the $5,000–$10,000 range. During 1978 there were 350 applications; in the first three months of 1979 they told us that 750 had come in.

Typical of such applications are a request from a Navajo community for a solar-powered wool-scouring unit; from a food cooperative in Massachusetts – currently importing heavily from California – to set up a solar greenhouse for year-round vegetable production, linked with a community cannery and garbage processing for compost; from a community garden group in California for a water-pumping windmill to be built and maintained by themselves; for a wood-fired bakery in Vermont; for an appropriate technology demonstration centre in Montana. The flow is endless; applications for grants totalling $7 million have already been received by the Centre, and its annual budget for this part of its work is less than $1 million.

The Centre has come in for some criticism for trying to do too many things at once: research, information, practical advice and distributing grants on a national basis. No doubt it will refine its operations as time goes on. But it has already proved that the kind of services it is offering are in demand in every corner of the country.[4]

People in Montana are very receptive to the idea of self-sufficiency and to encourage new ideas they have started a centre for innovation. This was set up by Jerry Plunkett, one of the people behind the National Centre for Appropriate Technology. Inventors, he argues, are at a disadvantage when dealing with large companies, and the idea of the centre for innovation is that through it, inventors can be funded to develop and commercialize their ideas. One centre has been started and within eighteen months it received 1,000 ideas; now it is planned to set up similar centres in the five states around Montana.

The staff of *Rain* magazine, another imaginative and well informed journal of appropriate technology, were our hosts in Portland, Oregon. *Rain*, originally intended to serve the A.T. communities of the Pacific north-west, now has an international reputation. Through them we were introduced to the Portland Recycling Team, the oldest and largest non-profit making recycling organization in the region. It was started in 1970 by one person with a pick-up truck; it now collects and recycles 800 tons of waste a month through several centres in Portland. Paper companies in the area buy different quantities of paper from them, and broken glass goes either to the local stained-glass firm or to a glass bottle manufacturer. They also recycle 200 gallons of car oil each month. They are virtually self-supporting, with an annual income of some $600,000.

Although they get most of their waste through contracts with public offices, the Team's main purpose is to stimulate and sustain community awareness of the need for conservation through recycling. To that end the team encourages people to separate their wastes, they encourage visits from schools to their centres, and they guarantee a regular and frequent waste collection service. The Team employ between sixty and 100 people, full- and part-time, throughout the year.

One of the young men who started the Recycling Team, Kevin Mulligan, now runs a commercial recycling consultancy unit. He and his partner told us that they can envisage a number of industries emerging from this kind of waste recycling – glass block making, aluminium ingot making and small-scale de-tinning. They can also see a considerable future for the small-scale manufacture of insulating material from waste: since transporting fluff is a costly business, they believe that any town of 30,000 people or over could set up its own insulation manufacturing unit, and the tax credits now being given for insulation will help to develop this industry.

From Portland we drove to Eugene. An attractive city full of trees and parks, it is the centre of a region that abounds in alternatives groups, mostly rural-based. The annual Oregon Country Fair usually brings together more than sixty organizations, including producer and consumer co-operatives, craft workers, environment groups, and others concerned with local self-reliance.

We started by visiting Cerro Gordo, a new town just started on a 1,200-acre site of Oregon woodland and meadow about twenty-five miles from Eugene. The town, and the community building it, have several unusual features. The town is being planned and built entirely by its future residents, 100 of whom live and work in neighbouring communities. They have formed a co-operative to buy the land. The houses are being built in groups that accommodate six or seven families, and will incorporate solar heating, a sewage disposal system linked with methane gas recovery, and rainwater collection systems. The community voted overwhelmingly to ban cars in the town, and in favour of bicycles, and a small trolley (train) system for public transport. They have set up a construction company, which will enable families to build all or part of their new homes. They envisage a town of about 2,000 people, relatively self-sufficient in terms of food, education and light manufacturing but relying, as a village should, on larger towns for other services, such as libraries and colleges. Meanwhile, in formulating their plans they encourage the participation of non-members through their quarterly journal, *Town Forum*, and by holding visitors' days.

On our way back to Eugene we stopped at Amity, a small foundation started about two years ago on a piece of waste ground in a poor quarter of the city. Bill Head, who runs it, told us that it was begun because of some very specific local problems. More than half of the local community comprise low-income families. Most of their houses have septic tanks, and these are now old and failing. Amity's aim is to develop low-cost ways of separating grey water (all the water used in a house for cleaning, washing and cooking) from sewerage water, and filtering the grey water for re-use, for example, for garden irrigation and fish tanks. Amity are experimenting with both greenhouse and small open-air fish tanks, raising species of fish which are familiar and acceptable to local people.

In their demonstration house there is a sand filtration system, developed by the Lane County Office of Appropriate Technology. This recovers the waste heat from the grey water before it is filtered and recycled, for use in flush toilets or in the garden, giving a 20–40 per cent water saving.

Nearly 1,000 local people had visited the Amity site during the six or seven months before our visit.

In Eugene itself there is the Lane County Office of Appropriate Technology (O.A.T.) which was, I believe, the first to be started by any county in the U.S.A. It was launched in 1978 and is housed in the council offices. It challenged the conventional view that no alternatives existed for sewage disposal, energy and recycling, and started to develop and demonstrate methods of grey water separation, solar heating and toilets that compost human waste. The Lane County O.A.T. also started the recovery from town waste of metal such as aluminium, brass and copper, which would otherwise have been incinerated. Their aim was to introduce such alternatives into the county's development plans as standard practice; and to that end they were working with the Department of Commerce, for example, to establish standards for composting toilets and the like. But they seem to have run into political problems not unconnected with the opposition of local contractors. The staff of the A.T. unit, at one time over a dozen, has now shrunk to one. We were told that the former staff have started a separate foundation to continue the work they began under the Lane County O.A.T.

In Eugene we had the good fortune to meet Lynn Miller, who farms eighty acres near Junction City and is the founder and editor of the *Small Farmers' Journal*. The journal, which he started after ten years in farming, is devoted to practical aspects of small farming and in particular, livestock rearing, horses and animal-drawn equipment. The journal is produced and edited by Lynn, his wife Christine, his father and two part-time helpers.

His own farm is an example of how small farming can be made to pay. He runs a mixed farm, with sheep, cattle, pigs and grows a variety of crops. All the horsepower on Lynn's farm is provided by horses. He believes that there is a great future for well designed, modern animal-drawn equipment.

He has recently formed the Small Farm Institute, with his farm as the demonstration centre, and for research and development work on small-scale and biological agriculture; it will also be a school for training those who want to start up on their own. Like the Organic Farmers and Growers in the U.K., he believes that there is much more to be gained by practical demonstration than by controlled experiments which lead to endless arguments among the agricultural 'experts'.

Lynn Miller's work is one aspect of what is now a nationwide revival of interest in the small family farm and rural communities based upon it. At the other end of the country, in Topsfield, Massachusetts, Elliott Coleman is running the recently formed Coolidge Centre for the Advancement of Agriculture. This is a 300-acre farm on which he intends to experiment with different systems appropriate to the New England environment. The basic intention is to find ways of farming efficiently by organic methods. He intends to discover whether, for example, a ten-cow dairy unit can be run economically if combined with cheese-making, and with the by-product, whey, being fed to pigs; and also what contribution can be made to the small farm by raising sheep, geese, chickens and quail on marginal land. Another part of the programme will be the testing and evaluation of small-scale farm equipment.

Practical work on small-farm and horticultural technology is also being done by, among others, the Rodale Press, Gardens for All, the Garden Way Manufacturing Company, *Tilth* (another farm-based journal), Ecotope, the Small Farm Management and Technology Project, the American Village Institute, the Land Institute, the Commonwealth of Massachusetts, and the Centre for Studies in Food Self-Sufficiency.[5]

One of the outstanding programmes of practical research on small-farm technology is the Small Farm Energy Project, a research and demonstration project sponsored by the Centre for Rural Affairs at Walthill, Nebraska. This is being designed by Roger Blobaum and Associates, a leading group in small-farm research in the U.S.A. They are working with twenty-four small farmers, who are given technical and practical assistance in adopting a wide range of energy-saving innovations. This help includes a small amount of cost sharing in buying equipment, and timber, hardware and other materials needed to construct and install energy-saving projects. (Another twenty-four similar farmers comprise the control group for comparative purposes; its only involvement is providing the project with economic and energy-use data.)

The project is designed to show how close this group of twenty-four low-income farmers can come, within three years, to making their farms self-sufficient in energy, without decreasing their total production. The technologies involved include wind-powered electricity generation and water-pumping; solar crop drying and solar heating of milking parlours and other livestock facilities; methane gas production for animal wastes; obtaining fertilizer from organic wastes from the farm, and from municipalities and food processing units; insulating farm buildings used for livestock; adopting minimum tillage methods, and farming without chemicals by means of crop rotations that include nitrogen-fixing legumes and the use of farm wastes.

All these energy-saving technologies, Roger Blobaum observes, are of proven value and some are part of a traditional farming technology. But they are innovations in the sense that they are not part of current standard management practices. Few have been tested on working family farms, and none have been tested for their effect on the net incomes of low-income farmers.

Completed projects already include solar hot water heaters on dairy barns, an attached solar greenhouse, three types of solar grain dryers, vertical wall solar collectors with storage, solar food dryers and other items. Most of these are home-built by local farmers using local materials. All this work is being closely monitored, and the Project is now publishing case studies giving technical and economic details of each innovation.

Biological husbandry is also attracting a growing number of practitioners and advocates in the U.S.A. The Rodale Press is very influential: *Organic Gardening* has a circulation of over a million. Recently they have branched out into the manufacture of equipment for the organic gardener, with their pedal-powered mechanical mule, and equipment for fish ponds and food processing. They have also established an organic gardening and farming research centre on a 300-acre site near Emmaus, Pennsylvania, to develop and test tools and equipment, and conduct practical research in cropping techniques and plants for the garden and the farm.

Countryside magazine, like *Tilth*, also practises what it advocates, on a 107-acre farm which is being converted from chemical to biological methods. Among several organic producers' associations, the Maine Organic Farmers and Growers Association is pre-eminent. It has done pioneer work in establishing standards and certifying organic farms.

An important piece of research was recently carried out into the comparative economic performance of organic and conventional farms in the mid-West. This was done by the Centre for the Biology of Natural Systems, of Washington University, Missouri. They compared the economic performance of fourteen organic and fourteen conventional farms between 1974 and 1976. Their chief findings were, first, that net income per hectare (two and a half acres) was the same for both types of farm; secondly, the organic farms used 60 per cent less energy to produce crops; and thirdly, that crop rotations on the organic farms reduced their soil loss by one-third compared with the soil loss on conventional farms.

Biological methods seem to be even better suited to smaller-scale operations, as John Jeavons showed us when we visited Ecology Action's experimental garden at Palo Alto, California, near the Stanford University campus. John, a political scientist and systems analyst turned horticulturalist, runs this three-quarter-acre garden with two assistants and a few apprentices.

His method derives from the French bio-dynamic system of

gardening, using deep, well aerated raised beds each measuring about six by twenty feet. Three tools are used – the u-bar, which is a long-handled fork with straight tines, employed to aerate (not invert) the soil; with this the average gardener can easily prepare a half-acre plot in a twenty-hour week. The second tool is a long-necked watering can of the kind popular in English gardens during the last century, which delivers water gently but faster than conventional sprays. The third is a miniature movable greenhouse made of wooden lathes and glass, which provides shade and protection from wind, and is easily shifted.

On his mini-farm, one-tenth of an acre, John Jeavons is growing potatoes, lettuce, onions, garlic, and is confident that in an eight-month growing season, and with forty hours a week work by one person, he can eventually reach a net annual income of some $20,000 from this plot. This kind of projection may seem exaggerated. But the results he is already getting are, to say the least, very impressive. He is now typically getting yields of between four and twenty times the Californian average. Compared with conventional methods, for each pound of food produced he uses one-half to one-sixteenth the average amount of water; one-half to one-sixteenth the average amount of nitrogenous (organic) fertilizer; and one-hundredth of the average energy expenditure, including transport to market. He uses no chemical fertilizers or pesticides.

Very detailed records of planting, feeding and production are kept, and periodically published by Ecology Action.

The biological control of pests – integrated pest management – is another 'technology' whose time has come. It too substitutes knowledge of natural systems for chemicals and capital equipment. Two specialists in this field are Bill and Helga Olkowski, who run the Centre for the Integration of Applied Sciences, a non-profit making research organization, near Berkeley. Several years ago, when they were working on waste management, they concluded that they had to find ways of getting rid of the toxicants being used to kill pests on trees in urban areas. They have since shown that this can be done by biological methods, using specific insect predators on specific pests. They now run an integrated pest control project for a number of cities in the region, including Davis, Berkeley and Palo Alto, which have nearly half a million trees in parks and streets. The results are a reduction of up to 90 per cent in the use of pesticides, with an attendant reduction in costs; fewer health hazards for the municipal horticultural workers; and reassurance for a public increasingly anxious to reduce chemical contamination of their environment.

A healthy sign of the growing strength and confidence of the groups and organizations working towards the renaissance of rural life, small farming and biological husbandry in the United States is their increasingly powerful and well organized representation in Washington. Rural America, a group which was originally concerned chiefly with rural housing, has now widened its interests to represent rural communities and family farms, and to promote alternative systems of farming and marketing that can liberate agriculture from its dependence on petrochemicals and fossil fuels. Another national organization which has recently entered the lists on behalf of the small farmer and the rural community is the National Conference on Alternative State and Local Public Policies. Their Agriculture, Land and Food Policy project, started in 1977, is aimed at influencing Federal policies by showing what has been done, and what more can be done, at State and local levels by way of helping family farms, preserving farmland, conserving energy in agriculture and increasing minority access to land ownership.

To return to our itinerary: from Eugene we headed for Sacramento, to the California Office of Appropriate Technology (O.A.T.). Appropriate technology has long been one of Governor Jerry Brown's interests, and the O.A.T. was started in mid-1976 by Executive Order. Today it has a staff of thirty and a budget of $900,000. From being an educational and knowledge centre, it is now providing a technical knowledge and technical assistance team of professionals operating both at government and local levels. Their aim is to make appropriate technology wholly non-partisan – to get it accepted as standard practice by government agencies, professional organizations and contractors.

The O.A.T. centre in downtown Sacramento encourages public access, and gets some 1,400 telephone calls and visitors, and 800 or so letters a month. Early on they started a Grants Newsletter to tell people how to apply for energy-saving grants that are available. They also produce a forty-page handbook intended for anyone investing in building work. This sets out to show that solar energy installations are easily within the reach of the average householder and illustrates the point with twenty-five examples from all over the State.

During our visit they were working on a handbook on alternative energy systems in agriculture, and another on the legal and technical aspects of starting community gardens. They already run a newsletter for community gardeners, as well as a more general newsheet on appropriate technology developments in California.

The O.A.T. have recently started a community services programme. Already several county O.A.T.s are being set up in California, and local groups are being helped to mobilize finance for small businesses, for example, or to start a community development corporation for making solar equipment. They have just launched a competition for the best invention in solid waste recycling that is applicable at community level. The community services that are available from the O.A.T. will shortly include expert advice on setting up co-operatives, and on fund-raising for community projects.

The O.A.T. energy programme includes the monitoring of a Federal grants scheme for inventors, aimed at encouraging alternative energy technologies. They considered 840 cases last year, and distributed $640,000 to some sixty individuals and firms. The O.A.T. view is that State, not Federal, awards should be made to inventors, since the State could then buy the patents and assign them to appropriate manufacturers.

Another project in this field is the setting up and operation of a State Energy Extension service, with funding assistance from the Department of Energy. Their aim is to build up an energy extension service parallel to the existing agricultural extension service, though perhaps less conventional in its thinking.

What struck me as particularly useful is their work on the training of solar engineering technicians. They are developing a six-month course for solar technicians, in consultation with universities, and with community development groups. The aim is to produce a standardized training 'package' which will give both employers and customers a guarantee of technical competence. This is particularly important in California, which has introduced a very generous tax credit (55 per cent) for solar installations (which can be spread over a number of years: if, say, the tax deductible amount is $1,700, and your State tax is $170, you pay no tax for ten years). One question they are now looking at is who controls the manufacture of solar equipment; they would like to see a mixture of large and small firms in this business, to avoid any possibility of monopolist price-rigging. They are also making a study of the various State agencies and departments, with an eye to encouraging them to use, or stimulate the use of, appropriate technologies.

The O.A.T. have a strong design team, which is incorporating appropriate technologies into new construction work. Thus they are encouraging the conversion of commercial greenhouses to solar energy. This is linked with the State re-afforestation programme, which provides a ready market for small nurseries growing tree seedlings. The team is currently evaluating some twenty-five different alternative sewage disposal systems; and they are funding a group at Berkeley to work on aquaculture, using waste water. The object is to establish a development centre for waste water treatment and aquaculture systems that could be operated at community level.

Other O.A.T. projects include a demonstration drought-tolerant garden in downtown Sacramento, and the compilation of a detailed catalogue of resource organizations that encourage and support local self-reliance; this is published as the Central Valley Atlas. Another example of their work which took our attention is a major – $170 million – programme for State buildings using solar energy. The first two buildings, designed by Sym van der Ryn, who started the O.A.T., are now running on passive solar energy systems, and use 65 per cent less energy than conventional structures, besides giving higher standards of comfort. The Capitol Bicycle programme is another typical instance of the O.A.T.'s imaginative approach. The O.A.T. distributes, loans, and services bicycles for State employees as an alternative to cars or taxis for inter-office transport. This has already saved tens of thousands of dollars, but that is incidental to the main purpose, which is to introduce acceptable energy-saving – and healthier – patterns of behaviour. We feel sure it would be safe to bet that the Governor's office in Sacramento is the only building in the Western world with a concrete stand by its entrance bearing the inscription 'Reserved for State Bicycles'.

En route to San Francisco we called in at Davis, which takes justifiable pride in being known as the energy-saving city.[6] The City Council's carefully considered reaction to the deteriorating city environment and rapidly rising energy costs was unorthodox;

they decided to invest in reducing energy demand – in ways that make the city a pleasanter place to live and work in. In the early 1970s they commissioned detailed studies of energy use, which showed that private cars represented 50 per cent of energy consumption, and space heating and cooling 25 per cent. So transport and building became focal points of their plan. They have worked out and applied a building code in which the location of buildings, and standard of insulation and construction, maximize conservation and the use of solar energy. They also hired consultants to design model solar buildings to show what could be done, and these are now being built for occupation by farm workers.

As we have already learnt, Davis is one of the cities that has cut back on its use of pesticides on the thousands of trees and shrubs that shade its streets. A highly efficient waste recycling unit, started on the university campus, now breaks even with sales of recycled paper, glass, and aluminium, totalling $3,000 a month.

The City's policy – revolutionary in the U.S.A. – is to curb the car and back the bicycle: its bike-ways, and bicycle safety measures, are unrivalled. As new streets are built, it is planned to reduce their width, on the grounds that this saves land, inhibits motor vehicles, uses less asphalt, and costs less.

Long journeys to work require transport and cause congestion. In most cities, zoning prohibits the development of workplaces in residential areas. Davis has passed an ordinance allowing home owners to operate small businesses, with sensible rules to preserve amenity.

Is it surprising that we felt that Davis is some years ahead of its time? It is a city of 33,000 people, and 25,000 bicycles. The City Council also run a fine public transport service of buses, which includes, to my delight, several ex-London red double-deckers.[7]

Davis University is one of the nine campuses of the University of California, and its Appropriate Technology unit runs the university's programme of A.T. research. Currently it is monitoring some forty projects. Many, as might be expected, are on various applications of solar energy, but many are also on organic agriculture and biological pest control. One project that caught our attention was on the Lorena adobe stove, now being widely introduced in Guatemala; another was on the design and construction of an energy-efficient refrigerator. The latter project, we were told, is in the Davis tradition, because it represents an investment in reducing demand, not increasing supply of energy. There is no doubt that as energy costs escalate, consumers will have to pay increasing attention to the life-cycle costs, i.e., initial, running and maintenance costs, of the hardware they use, and not only its initial costs. The variety of experiments going on, both inside the university and in the city of Davis, may help this process along.

The Davis city experiment shows how it is possible to save energy and improve the environment by the intelligent use of existing technology. At the Farallones Institute centres, which we next visited, we saw some of the new generation of life-support systems in the making.

The Farallones Institute was started in 1974. It is an alliance of scientists, engineers, architects and artisans, whose purpose is to explore and evaluate technologies and systems based on the use of renewable resources to meet the basic human needs of food, shelter and energy. This work is done in two centres, to show the application of these systems in both rural and urban environments. The rural centre is on the site of an eighty-acre farm north of San Francisco. It has a permanent staff of eight, and about a dozen apprentices and volunteers. The centre's activities represent a combination of action-learning with action-research. Staff and students grow most of their own food in well tended organic gardens and orchards, live in energy-efficient dwellings built by themselves, learn crafts such as blacksmithy and generally follow a lifestyle which is in harmony with natural systems and eliminates waste of resources to the maximum possible extent. There is also an experimental and research element in all the work done. Thus the centre is making comparative studies of several passive home heating systems and different types of solar greenhouse, and of biological pest management techniques, the results of which are published in bulletins and fact sheets.

The urban centre comprises a house and garden in an industrial-residential neighbourhood in Berkeley. It also demonstrates tested technologies and experiments with new ones, through classes and workshops, and somehow manages to cope with tens of thousands of visitors a year. The emphasis of the urban house is on systems and methods that are within the reach of the average householder, such as grey water filtration and heat recovery, solar heating, insulation, organic gardening, aquaculture, food preservation and the like.

A growing number of young people are coming to the Farallones Institute as part of their degree course with the Antioch University environmental studies programme. This imaginative degree course, which covers various aspects of appropriate technology such as solar energy and design, environmental horticulture, and ecosystem management, gives academic credits for on-the-job experience at centres such as Farallones.

Another of the highlights of our visit to San Francisco was a meeting with David Olsen, head of the New School for Democratic Management. This school was launched two years ago as a project of the Foundation for National Progress, an employee-controlled educational corporation which is probably best known as the publisher of the counter-culture *Mother Jones* magazine (for the benefit of non-American readers, *Mother Jones* is a splendid combination of the qualities of the *Sunday Times* Insight team, *Vole* magazine and *Private Eye*). The school challenges the conventional view that democracy and good business management do not mix. It is committed to bringing about economic democracy through decentralized forms of entrepreneurship and ownership.

David Olsen is a former Tufts and Cornell professor. He started the school, he says, after working in a research collective that collapsed owing to poor management. He estimates that there are now more than 10,000 co-operatives, alternative enterprises and worker-owned and managed firms in the U.S.A. He sees these as the basis of a restructured economy comprising a wide variety of neighbourhood production units – worker-owned and democratically controlled.

Alternative organizations of this kind, however, are notoriously lacking in essential management and business skills. The missing factor, David Olsen decided, was that there existed no body of theory or sound business practice for running a business, without all the trappings of corporate structure, but which recognized the unique problems arising out of smallness and democratic control. The object of his school is to make good these deficiencies. Both the form and the content of the New School's courses are specially tailored for the small community-based organization. The courses he runs are short – one or two weeks – but intensive, because managers of small firms cannot afford to be away for long. For this reason too the School is mobile; its classrooms are in San Francisco, but it has also run courses in cities all over the country. Since 1976, more than 600 people representing some 450 organizations have attended the school.

The curriculum, besides covering traditional topics such as starting a business, negotiating leases and contracts, and handling data, also includes classes on special problems that beset community enterprises: how to run meetings, resolve conflicts, make agreements with and among workers. We noticed with pleasure that one of the organizations that had been represented on a course had the charming name of The Cheese Rustlers of Minneapolis. We had met before: they are an all-women collective that shares warehouse premises with DANCE, and rustles up cheese very efficiently for the Minneapolis co-operatives.

The New School for Democratic Management represents one of the growth centres, so to speak, of the worker management movement in the U.S.A. While it concentrates on one essential function, training, other organizations exist to set up and promote worker-owned firms and co-operatives.

Two of the outstanding organizations in this field are the Delaware Valley Federation for Economic Democracy, and the Industrial Co-operative Association. Both grew out of the Federation for Economic Democracy, a national body which ceased to exist when it was agreed that the prime need was to build up smaller local organizations with their own staff and projects.

The Delaware Valley F.E.D. is in Philadelphia. With foundation support it is carrying out a feasibility study of the reopening, under democratic ownership, of a factory threatened with closure. It is also starting a credit union and a community development corporation, and running study groups on workplace democracy. It has been very active, along with other groups, in supporting two congressional bills of great importance.

One is the Co-operative Bank Bill, which provides a major new source of funds chiefly for consumer co-operatives, but also for housing and producer co-operatives. This bill is now law, but support is needed to ensure that its rules and organization will allow the funds to be used creatively. The other is a Voluntary Job Preservation and Community Stabilization Bill, also now law, developed by William F. Whyte along with several Congressmen. The Act provides for the use of some $20 million of employee unemployment insurance to buy shut-down plants for employee ownership.

The second organization that grew out of the late Federation is the Industrial Co-operative Association (I.C.A.), which has its offices in Cambridge, Massachusetts. It has developed two worker co-operatives: the International Poultry Plant, a poultry processing co-operative in Willimantic, Connecticut, and the Colonial Co-operative Press, a typesetting and printing company in Clinton, Massachusetts. The I.C.A. has obtained several hundred thousand dollars for each enterprise, and in both cases has helped the workers develop the co-operatives from the planning stage to commercial operation. The poultry plant employed fifteen people in 1979, and it was hoped within a year to employ fifty. The printing plant employed twenty, mostly in typesetting. It hoped to employ sixty within a year, and up to 200 within a couple of years.

The I.C.A. is currently working on several other projects, including a wool spinning factory in Maine, and a meat processing plant in Boston.

The Association has three full-time and two part-time staff, and its Board includes representatives from local minorities, and from several of the co-operatives it has helped to establish.

From San Francisco we started on our return journey, which took us first to Denver, Colorado, to the Domestic Technology Institute. This is run by Malcolm Lilleywhite, who left the aerospace industry to devote his skills to self-help and community technology. The Institute has developed a broad capability for research and training in the field of small-scale technology, especially in energy conservation and economic development through the application of appropriate technology at the community and small business level.

One of the Institute's projects, which has been going for two years, was the construction, along with people in the community, of a 5,000 square foot greenhouse in Cheyenne. This employs one full-time horticulturalist, and many elderly people as volunteers. Along with this there is a methane plant using greenhouse and domestic wastes. The output of the greenhouse is being carefully measured and recorded, so that other communities can be encouraged to undertake similar projects. Another community programme with which the Institute is closely involved is in San Luis valley in south-central Colorado. There are few gas lines in that area, and the chief fuels are propane and wood. Because shortages of both are imminent, people are taking readily to using solar energy. The Institute helped to promote the San Luis Valley Solar Energy Association, which has built hundreds of adobe houses with passive solar energy systems.

Among their research and development projects there is one on a solar-assisted processing system for making ethanol fuel. This is a fermentation and distillation unit for processing agricultural waste into alcohol for fuel, and thereby providing new markets for local farmers and work for the unemployed. They are now working on a pilot project that will make about 250 gallons of alcohol a day. So as not to compete with food crops it will use spent grain and apple waste as raw material. Because large-scale plants of this kind would compete unavoidably with food production, a number of small-scale, community level plants are envisaged, each covering about a thirty-mile radius. Besides producing alcohol such plants could also make animal feedstuff and methane gas.

Solar energy is a very rapidly expanding technology in this part of the world. In 1977 the federally founded Solar Energy Research Institute was started, managed and operated by the Mid-West Research Institute at Golden, Colorado. It now has a staff of some 300 people. Farther south, New Mexico can still claim to be the most solar-powered State in the Union. It has under one half of one per cent of the U.S. population, but accounts for 10 per cent of all solar energy installations. It seems likely to retain this lead, owing to the efforts of the New Mexico Solar Energy Association, which has promoted a steady flow of pro-solar legislation in that State during recent years; and the *N.M.S.E.A. Bulletin* is a model of both do-it-yourself and scientific information on the subject. New Mexico's pre-eminence in this field owes much to Peter van Dresser, an engineer and architect who has worked on passive solar energy systems for more than twenty years, and has been advocating and helping to bring about small-scale, self-sufficient communities for the past quarter of a century.

Denver is also the centre of a flourishing community gardens movement. We met one of its veterans, Jim Fowler, who started a Centre for Biological Self-Sufficiency about three years ago. On a four and a half acre plot on the outskirts of the city food is produced for the Centre workers and for sale; about forty people work the plot collectively, each drawing out produce according to his or her contribution. Marketing is done through some forty stores in the city. They are now proposing to set up a small oil press to process their sorghum crop. They are devoted organic gardeners, and nearly one acre of their plot is used for composting. Jim Fowler estimates that about half of the city's 150 public vegetable gardens have stopped using chemicals.

Our next stop was at a much bigger collective. This was The Farm, near Nashville in Tennessee, which is basically a religious community, started in 1971 by Steve Gaskin and several hundreds of his followers. Its members take a vow of poverty, and try to pursue non-violence in all aspects of life. Now numbering about 1,500, they live in a self-built village which is widely scattered over a large and heavily wooded property owned by them. The development of the community has not followed any grand design. They learnt by experience, they told us, and met communal needs as they arose. They now run a health clinic staffed by two doctors, with nurses and midwives. They also run a food preparation unit, a corn grindery, a cold store, a food distribution centre, a printing shop, a laundromat and a community broadcasting (C.B.) radio station.

An offshoot of The Farm, occupying the same premises, is Plenty, a foundation set up to help other communities at home and overseas. This was started in 1974, originally with the intention of giving food to people poorer than themselves. When an earthquake hit Guatemala in the mid-1970s, they sent their first team overseas. There are now about twenty people from The Farm working in Guatemala, helping to rebuild communities. More recently they have sent a C.B. radio team to help with a paramedical programme in Bangladesh: to provide it with a low-cost communications system. Plenty is a tax-deductible foundation, while The Farm itself is really a latter-day monastery.

The Farm is a strictly vegetarian community and relatively self-sufficient in food. Their main source of protein is soya. We were shown round the soya preparation plant which contained various pieces of equipment that they had made or adapted for themselves: the hydraulic press for extracting soya milk, for example, had started life as a machine in a toothpaste factory. From this unit, which produces 250 gallons of soya milk a day, people come to collect soya milk, bean curds and soya ice cream, which is a great favourite with the children and, as our guides explained, the best way of ensuring that the children get enough protein.

As we walked around the village we watched the construction of a new communal meeting place, a large dome capable of seating 2,000 people. This was being built by their own craftsmen. And a new solar-heated school, for several hundred children, was nearing completion. When they are not thus engaged The Farm craftsmen are doing contract jobs in surrounding towns and villages, and their earnings are The Farm's chief source of cash income – not that anyone needs much cash, as food and services are provided free to all.

The C.B. radio station keeps The Farm in touch with their team in Guatemala, and with other groups working in the Bronx, in California and Chicago. They now intend to start making their own clothes, and for this purpose a Guatemalan weaver has been invited to stay with them and teach them to weave.

The Farm's governing council, of about twenty people, is appointed by consensus. Much of the administration and planning is devolved to committees – on town planning and building, the clinic, farming, food preparation and so on. Each group is responsible for recruiting volunteers for particular tasks. Government and administration at The Farm do not seem to present many problems.

In contrast with The Farm, most poor communities are anything but self-reliant, and urban slums reduce their inhabitants' self-confidence to vanishing point. In Washington, D.C., our next stop, we visited a small organization that has taken up this challenge, the Institute for Local Self-Reliance (I.L.S.R.). It was started in 1974 by David Morris and other community workers, who were convinced that if community development was to be anything more than the delivery of services from above, people needed to get access to technologies that they could handle, and to be involved in productive work that would contribute to their community.

Today the I.L.S.R. is a group of fifteen specialists – in community organization, energy, waste utilization – that offers practical knowledge and technical assistance to community groups in high-density areas, generally inner cities.

Their guiding principle is to marshal local materials and local production for local use. When they are asked for help by a community group, their approach is to start by identifying what resources the people have available, such as land, buildings, skills, materials; and next they suggest alternative ways in which these resources might be combined into productive activities. Depending upon what people decide, the Institute provides appropriate on-the-job technical and organizational assistance and training.

The success of this approach is well illustrated by a project in South Bronx in New York, on which the I.L.S.R. have been working for three years. South Bronx has many vacant lots – to a Londoner it looks like parts of the East End after the bombing of the Second World War – and it produces, or has access to, a lot of garbage. To make the derelict lots usable again, as gardens or recreation parks, a minimum of eight inches of new topsoil is needed; but topsoil in New York sells for about $13 a cubic yard. The Institute put these facts together and came up with a proposal for a large-scale compost-making operation to be run by the local community. Eighteen months and a great deal of hard work later the Bronx Frontier Corporation was running an enterprise that produces over 1,000 cubic yards of compost a month. What was formerly a cost to the taxpayer (waste disposal) has been turned round to become a revenue-earning project of the community and a means of reclaiming land. This in turn has sparked off other community initiatives and local groups are now being formed to run community gardens and self-help housing co-operatives, and one has asked for help in erecting a windmill generator. The Institute also sees housing improvement as a means of creating new skills and incomes, and leading to greater local self-reliance. It is helping the residents of tenement blocks – many of which have been abandoned by their former owners – to refurbish them completely and run them as co-operatives. The skills acquired in the process will enable local people to set up in business on their own, provided they can get some help to begin with.

Altogether the Institute is working with more than fifty communities, all over the country, on various aspects of waste recycling, energy conservation and urban gardening. They also do contract work that supports their objectives. Under a contract from the Community Services Administration they have completed a detailed feasibility study for community-based cellulose insulation factories, and provide a back-up service of technical assistance to community organizations interested in going into the cellulose business. They are also working on a study of the potential for energy self-reliance in Washington, D.C. This includes an assessment not only of the technical possibilities for energy conservation and the use of renewable energy sources, but also of organizational aspects – the tools and authority available to the district to implement a policy of energy self-reliance.

A similar vision of decentralized economic and political power, based on technologies that are non-violent towards people and other living things, informs the Movement for a New Society. Visiting one of its centres in Philadelphia, we learnt that their emphasis is on people rather than hardware. As a network of small autonomous groups working for radical non-violent change, it represents a valuable human resource for a wide range of alternative groups and counter-culture movements in all parts of the country.

Small groups, or collectives, are the basic units in the network. Several different kinds of collective have evolved. There are action collectives whose members carry out non-violent direct action projects, on such issues as the dangers of nuclear power and the need to promote alternatives; militarism; and women's rights. Another category is 'alternatives' collectives, whose members involve themselves in creating, serving and consuming without exploitation: groups practising alternative forms of education, health care, food co-operatives, communal gardens, recycling centres, home repair teams. A third category are called life centres, larger communities whose purpose is educational, that of training and supporting effective social change agents. Living in such a community involves taking part in training workshops on skills exchange and community organization, participating in direct action projects, and following a non-violent lifestyle. The life centre in Philadelphia, for instance, is one such action training community of about 130 people in twenty houses located in a west Philadelphia neighbourhood.

In short, the network represents many facets of the growing revolt against giantism in economic and political life, violence, and consumerism. Recognizing that any real change in contemporary industrial society must start with the individual, with changes in personal beliefs and lifestyles, its members are trying to live now as they envisage the future should be. But this by no means implies a withdrawal from action and commitment, and very many alternative technology groups throughout the country would be the poorer if it did not exist.

The Movement for a New Society now has thousands of members and some thirty regional centres in the U.S.A.; and it has links with similar organizations in India, Japan and the Netherlands.

Our tour ended in New York, at the People's Development Corporation (P.D.C.) in the Bronx. The P.D.C. was formed in 1974, as a community organization pledged to revitalizing its neighbourhood through self-help techniques in housing renovation, job training and economic development. Membership – which is open to anyone who contributes work to community projects – has risen recently from thirty to over 200.

During our visit, four six-storey tenement buildings were being rehabilitated, with funding by the city and a number of banks; and people from the community were investing their labour (sweat equity). The first completed block now houses sixty people and is run as a co-operative. All hot water for the block is solar heated by rooftop solar collectors. Also on the roof there is an experimental fish tank, while the basement houses a compost making unit. One million worms speed the process and the compost goes on the community garden.

In the rehabilitation of the buildings, the work of each trade – carpentry, plumbing and so on – is done by a team, which as it gains experience can form itself into a small company or co-operative and take on outside work. One group has already started a cabinet-maker's workshop in a nearby block.

In five other buildings in the neighbourhood, tenants' associations now manage the buildings they live in. Two years ago these buildings were on the brink of being abandoned. The landlords were not providing heating or hot water, and in desperation tenants were leaving. Now all the buildings have a heating and hot water system and are fully occupied.

A growing number of community groups in New York are now taking up self-help housing, or urban homesteading, as they call it. They are advised by the Urban Housing Assistance Board. One group that started in this way is the East Eleventh Street Movement, an active neighbourhood organization with a programme similar to that of the P.D.C. In both instances there is a clear recognition that a programme for improved housing should be part of a wider plan of community economic development and self-reliance, and they are showing how it can be done.

The P.D.C. are now starting work on several open-air projects, including an urban garden, a recreation ground and a children's playground. One of their more ambitious schemes is to construct a large greenhouse, to house an integrated system of food production including vegetables, fish, earthworms, rabbits and poultry. They also plan to start a laundromat, a pharmacy, a restaurant and a bakery; a materials' recycling warehouse and a construction co-operative; and an energy unit specializing in solar installations and boiler maintenance.

There could not have been a more rewarding end to our 6,000 mile tour than to see the technical competence and determination of the young people who run the People's Development Corporation.

In an economy which, above all else, has been built on the prodigal use and misuse of resources, it would be surprising if the need for a change in direction, towards a conserver society, were to pervade the national consciousness quickly and without resistance. What strikes an outside observer about the U.S.A. is the extent and variety of the groups and organizations that are already advocating this kind of change and are actively involved in making it happen.

This readiness to explore alternatives is not restricted, as it is in Britain, to the non-government sector. Two official reports published in the U.S.A. in mid-1980 illustrate this. They could have a considerable effect in changing government and private sector policies and accelerating the development of non-violent and sustainable technologies.

One is the *Report and Recommendations on Organic Farming*, prepared by a U.S. Department of Agriculture team appointed by the Secretary of Agriculture.[8] The report reflects a growing concern about the costs and ill-effects of conventional chemical-intensive farming, and a growing interest in a system that avoids the use of synthetic fertilizers, pesticides, growth-regulating chemicals and food additives. Three of the things the team discovered were, firstly, how little is really known about organic farming, its extent and practices; secondly, and contrary to popular belief, that 'most organic farmers have not regressed to agriculture as it was practised in the 1930s'. They make full use of modern machinery, crop varieties and so on; and thirdly, that well established organic farms recorded crop yields similar to nearby chemical-intensive farms. The report includes a series of specific recommendations for research and development, education and extension, and for building organic farming into U.S. government policies and programmes.

The other report deals with world-wide issues. It is entitled *Global 2000*, and was prepared by the U.S. Council on Environmental Quality and the Department of State.[9] The study was commissioned by President Carter in 1977, as a foundation for U.S. government long-term planning. Essentially the report takes a realistic look at what will happen to the world's population, resources and environment if present policies of development, and their supporting technologies, continue.

The central conclusion is that present policies can lead only to a progressive degradation and impoverishment of the earth's natural resource base. As might be expected, the poorest countries will be the hardest hit, being likely to lose 40 per cent of their forest cover by the end of the century, and to suffer shortages of energy, water and food on an increasing scale. On present trends the report forecasts too a growing impact on the planet's atmosphere by industrial pollution, carbon dioxide build-up, and destruction of the ozone layer. It also points to the growing risks of radioactive contamination of the environment, especially from the disposal of radioactive wastes from nuclear power stations. Some of these by-products of reactors, it notes, have half-lives approximately five times as long as the period of recorded history.

In large measure this report amounts to an indictment of the conventional trend of technology in industry and agriculture. The authors recognize that the necessary changes go beyond the capability of any one nation. 'But our nation itself can take important and exemplary steps. Because of our preeminent position as a producer and consumer of food and energy, our efforts to conserve soil, farmlands and energy resources are of global, as well as national, importance. We can avoid polluting our own environment, and we must take good care not to pollute the global environment.'

While studies are not action, they are an essential first step. And both these studies suggest that policies favouring more appropriate technologies are beginning to emerge in the United States.

6

CANADIAN INITIATIVES

Among the rich countries, Canada presents a unique combination of development problems. It is a major producer and exporter of primary products, renewable and non-renewable; and a significant part of Canadian industry is externally owned and controlled. In these respects Canada's problems are more akin to those of a typical developing country, which has complete political independence but is still a victim of a kind of colonial exploitation. Seen from the standpoint of many Indian and Inuit (Eskimo) communities, however, or indeed that of many other parts of rural Canada, the centre – federal government and metropolitan Canada – looks much more like a colonial power than anything else. A keen appreciation of this subservient economic status is at least one of the forces behind the Quebecois movement. Though it is less forcefully expressed, there is the same understanding in the Maritimes and in the Canadian 'middle north', roughly the belt of territory bounded by the prairies to the south and the tundra to the north. It was from Newfoundland that a demand came as early as 1968 for the development of technologies that are more in harmony with the needs and resources of people living in these areas and could arrest the disintegration of their communities.

The same theme was taken up by the Canadian Council on Rural Development, an officially sponsored body which, in a series of meetings held in the mid-1970s, strongly advocated the introduction of self-help technologies for rural communities in Canada's mid-north.[1] These ideas also informed the government of Manitoba under Premier Schrier. Between 1975 and 1978 they drew up a plan for the development of northern Manitoba. Their strategy was that of

> converging local resource use with local demands and needs – to cut down on the costly and unnecessary two-way flow of goods and to build up a strong northern economy with extensive backwards and forwards linkages and economic activities which maximize local value added . . . the scale of investment opportunities should, as far as possible, be such as to facilitate local ownership and control at the community level . . . the basic emphasis of the strategy will be on productive investments in and for the North rather than on welfare and other transfer payments.[2]

The plan set out how these principles would apply to forestry and fishing, agriculture, housing, consumer goods, energy supplies and education. Regrettably these plans were not pursued by a new government which took over in Manitoba in 1978. As a strategy for rural development in Canada they would be hard to improve upon.

Another Canadian initiative of great value was the publication in September 1977 of the report *Canada as a Conserver Society: Resource Uncertainties and the Need for New Technologies*.[3] This was prepared by a committee of the Science Council of Canada, headed by Ursula Franklin, Professor of Metallurgy and Materials Science, University of Toronto. In their view

> a Conserver Society is on principle against waste and pollution. Therefore it is a society which
>
> > promotes economy of design of all systems, i.e. 'doing more with less'; favours re-use and recycling and, wherever possible, reduction at source;
> > questions the ever-growing per capita demand for consumer goods, artificially encouraged by modern marketing techniques, and recognizes that a diversity of solutions in many systems, such as energy and transportation, might in effect increase their overall economy, stability, and resiliency.
>
> In a conserver society, the pricing mechanism should reflect, not just the private cost, but as much as possible the total cost to

society, including energy and materials used, ecological impact and social considerations. This will permit the market system to allocate resources in a manner that more closely reflects social needs, both immediate and long term.

This is a report of outstanding quality, because – rather like *Small is Beautiful* – it not only identifies the inherent contradictions and dangers of conventional industrialization and its support systems, but goes on to spell out the new opportunities for invention, initiative, and employment that would be opened up by a conserver society which put net human benefit before gross national product. Such opportunities arise, the authors of the report point out, at every turn: in energy conservation and the use of renewable forms of energy, and in the use made of renewable and non-renewable resources, there would be scope for many new technologies, new business enterprises and more creative work. The report includes practical recommendations that could be acted upon immediately – by Canada or any other industrial country – in the fields of transport, shelter and community, renewable energy, materials conservation, and industry and employment. A conserver society, that is, means much more than just conservation.

Conservation becomes a function or output which arises both from an understanding of our system and its various subsystems, and when that understanding leads to innovation. In other words, conservation occurs as the result of a much more appropriate and sensitive reordering and redesigning of activities. This is the creative core of the Conserver Society concept, and it is closely linked to the need for diversity and flexibility.[4]

This constructive view of Canada as a conserver society has evoked an enthusiastic response from local groups and individuals throughout the country but not, so far, from the centre of power in government and industry. In Canada, as in Britain, the action is with voluntary groups and organizations, with some support from universities and colleges and the lower echelons of government; but with the central establishment still clinging to the notion that 'normality' – the limitless growth mania of the 1950s and 1960s –

is the goal, and conventional industrialization the means of achieving it. For this reason, this excellent report has not had the widespread publicity it deserves.

The advocates of the conserver society in Canada have a long haul ahead of them. But their constituency is growing. The Berger report on the Mackenzie Valley pipeline enquiry gave people in Canada – and far beyond it – a sense of the values that put people before big business. In this report Justice Thomas Berger recommended that no pipeline should ever be built across the northern Yukon; and that any pipeline down the Mackenzie Valley should be postponed for ten years. He stated:

> The risk is in Canada. The urgency is in the United States . . . How much energy does it take to run the industrial machine? Where must the energy come from? And what happens to the people who live in the path of the machine?
>
> Native people have been told that they cannot compromise, they must become industrial workers or go naked back to the bush . . . Yet many of them refuse . . . They see that complete dependence on the industrial system entails a future that has no place for the values they cherish . . .
>
> We have never had to determine what is the most intelligent use to make of our resources . . . Will we continue, driven by technology and egregious patterns of consumption, to deplete our energy resources wherever we find them? . . . without asking Canadians to consider alternatives. Such a course is not necessary, nor is it acceptable.[5]

In asking what kind of future Canadians wanted for their native people, Judge Berger was really asking what kind of future Canadians wanted for themselves.

The Canadian Federation of Independent Businesses is a powerful advocate of small-scale technology and decentralization. The Federation argues for more knowledge and skill-intensive manufacturing; new rural industry; more independent retailing, and more decentralization of government. To support their case the Federation commissioned Dr Rein Peterson of the Faculty of Administrative Studies, York University, Toronto, to make a study of the legislative and other initiatives that must be taken to build a new economy for Canada. In what has become a classic work[6] on small business – he draws on international experience in this field – Dr Peterson argues that:

> Cultural goals and economic targets cannot be as easily separated as traditional economists would have us believe. If society as a whole is to provide a positive, alternative response to the essentially negative possibilities [of conventional, large-scale industrialization] . . . we must ensure that our governments pursue a goal of the fair dispersion of wealth and power, and not income. Fixation on income encourages too many of us to become employees, dependent ultimately on 'someone else' . . . Widespread acceptance of the goal of dispersion of wealth and power would lead to greater individual freedom, have a snowballing effect on Canadian creativity, and provide opportunities for a far larger proportion of us to become self-sufficient.

Dr Peterson argues that separate industrial policies should be developed for the big, foreign-owned, and the small business sectors in Canada. Policy on the small sector would be based on an explicit recognition of the small business as a centre of innovation and self-reliance, and on the need to promote regional self-sufficiency through appropriate-scale technology.

Another organization of growing influence which supports the values of a conserver society from a different standpoint, is the Vanier Institute of the Family. The established tendency of the conventional economy, argues the Institute's director, Bill Dyson, is to regard families and communities in terms of their capacity to consume, without question, an increasing volume of goods and services and to accumulate long-term debt in the process. One of the Institute's objectives is to bring about a new perception of the family and the community as the true economic foundations of the economy, and to strengthen family and community self-reliance.

Like James Robertson in Britain and Hazel Henderson in the United States, the Vanier Institute maintains that the conventional economic calculus of the cash economy ignores the vast amount of useful and productive work that is done by families and within communities. The fact that it is unrecorded by national income or employment statistics makes it no less real; and the more that families and communities can become self-sufficient – the more we can 'de-institutionalize' economic activity – the less dependent we shall be on the large-scale conventional structures, and the less vulnerable to the consequences of their progressive collapse.

The Vanier Institute, in the course of promoting this wider understanding of the indispensable economic role of family and community, sees a distinct

> new trend by many people to shift activities, which had been sucked up into the economic superstructure, back down into the community and the individual household. These activities include: the growing rejection of job transfers, promotion and mobility, because home and community stability are preferred; the vast increase in home-food production and preserving, and organic gardening [a movement which owes much to the excellent *Harrowsmith* journal]; a wide array of conservation and conserving activities; rapid growth of hand-crafted goods and personally and group-provided services; the spread of craft and do-it-yourself learning; alternative housing; home heat and home energy production; small grocery co-operatives; urban and rural group-living arrangements, and the spread of the voluntary simplicity movement.[7]

Moving from the realm of advocacy to the field of action, the two most striking and imaginative programmes I have encountered in Canada are those of Prince Edward Island, and Sudbury, Ontario.

The economy of Prince Edward Island provides a classic example of the 'hinterland effect'; the destructive impact of large-scale technology upon the structure of a community that is on the periphery of a centralized metropolitan economy. Within the past fifty years P.E.I. has changed from a virtually self-sufficient agricultural province to one which is almost totally dependent on the outside world.

Only thirty years ago there were some 11,000 farms on the island, mostly small mixed farms. Every village was self-sufficient in the basic necessities of a rural community – blacksmithy, carpentry, milling and baking, leatherworking and so on. Today these skills, which give variety to an agricultural community, have gone, and there are fewer than 3,000 farmers left, mostly growing potatoes. Monoculture and mechanization have made them dependent on imported energy and imported manufactures. Commuters have replaced the local tradespeople. The local tradespeople have become commuters to city jobs.

Most of the farm machinery we use is made in the United States. Our gasoline and oil come from Venezuela. The nitrogen component of our fertilizers is made with natural gas from western Canada. There are supplies of potash rock in the region, but phosphates present another story. Our supplies come from Florida. But planned production cuts there in the next ten years will make us reliant on Morocco for phosphates. The active ingredients in our petroleum-based pesticides are not made in Canada; we depend entirely on the U.S. and Western Europe.

Monoculture and mechanization have also jeopardized P.E.I.'s land resource. Our soils are an exceptionally light sandy loam, high in acidity, low in natural fertility, and easily erodable. The provincial soil laboratory reports that each year a few more soil samples come in with organic matter content that is not detectable by normal methods of soil analysis . . . tons of Island soil are washed and blown away each year.

Island farmers are also increasingly dependent on a single cash crop – potatoes. Export sales of potatoes can account for 50% or more of gross provincial farm income . . . Wild price fluctuations from year to year tend to drive smaller producers out of business. Concentration on potatoes has other ominous effects. Seed potato export markets demand frequent, high rates of pesticide application. Farms have increased in size, and land values increase. Larger machinery is demanded, in fact the largest potato harvesting machine in the world is now used on P.E.I. . . .

The potato economy depends in turn on availability of cheap transportation. Most of our tablestock potatoes are sold in Quebec or Ontario, where the cost of production is not very different from that on the Island; a transport subsidy make P.E.I. potatoes competitive. If rail rates rise, it will be increasingly difficult for the Island farmer to compete in his principal domestic markets. I find it rather insane to ship a product like potatoes that are 80% water from one end of the country to the other when they can be grown competitively everywhere.[8]

In all these respects P.E.I. is no more unfortunate, perhaps, than other parts of the Maritime provinces . . . or elsewhere in the Canadian hinterland. Where they were singularly fortunate was in having a premier, Alex Campbell, who recognized the imminent dangers for his province of a growing shortage of oil and other resources. Instead of becoming increasingly dependent upon an obviously highly vulnerable system, the government of P.E.I. decided to investigate the feasibility of basing their development on an intelligent use of local resources to meet local needs.

In 1975 the Institute for Man and Resources was set up as an independent, non-profit making organization, with a mandate to:

Develop, test and assess appropriate energy systems based on solar, wind, water biomass and other sources;

Identify potential benefits from local manufacture of equipment related to these systems;

Develop, test and assess appropriate food and crop production, transportation, and shelter systems which will reduce use of non-renewable resources and inappropriate equipment;

Provide service and assistance to individuals, groups, communities and governments in making appropriate resource management choices;

Test and evaluate new inventions for the better use of renewable resources;

Gather and distribute information on appropriate food, shelter and energy systems; and

Design and implement programmes and processes beneficial to the long-term economy of Prince Edward Island and to other areas of the world.

Three years later, with a staff of twelve, the Institute was well into a comprehensive programme covering renewable energy – wood, wind, solar, mini-hydro; energy management; agriculture, aquaculture and fisheries; and consultation and policy advice.

Their work on renewable energy includes the development of a sustainable yield wood-fuel supply system and distribution facilities. This programme includes test installations of advanced wood-fired heating systems. They are also exploring the potential for a wood-fired district heating scheme, a wood-fired gasifier and other related systems.

Wind energy sources are being evaluated, with a view to the possibility of the P.E.I. grid system absorbing wind-generated electricity, and also with the aim of developing non-electrical uses of wind. Solar energy is being explored in depth, both for space heating and for domestic hot water. The Institute has collaborated with a local developer to incorporate an air-based, rock-storage solar space heating system in an apartment building, and with Island home owners on solar assisted hot water units.

Low-head hydro-power has limited possibilities on P.E.I.; the Institute has identified seven potential sites, and it has sponsored a project to restore a dam and a hydro-power complex on the Island.

Another of the Institute's major programmes is on energy conservation. They have developed a list of energy conserving measures for the home owner, showing costs and savings, and they are carrying out detailed tests on building materials and methods. Energy audits have been made on a variety of farms on the Island with the object of designing appropriate conservation methods and renewable energy alternatives on the farm.

Agriculture and fishery also offer scope for developing systems that are ecologically sustainable and which reduce dependence on non-renewable resources. This programme includes a study of locally available fertilizers, and soil additives such as mussel mud and seaweeds. Experiments with a salmon and trout hatchery using recirculating water and solar heating have shown very promising results.

The Institute has not neglected two of the supporting services that are essential if appropriate technology is to become standard practice throughout a community. One is a consultancy service. The Institute's energy experts have already been called in by several island communities, store owners, individuals and contractors, and design advice has been given to small firms interested in manufacturing and marketing a low-cost hydraulic wood-splitter, air-tight woodburning stoves, and commercial scale woodburning furnaces.

The second supporting service is on legislative and policy

measures. The Institute has, for example, drafted a series of legislative and policy proposals which promote high energy management standards, and facilitate the introduction of renewable energy into the provincial economy; these include solar rights legislation, and incentives for re-insulation.

One part of the Institute's programme that has attracted most publicity is the Ark. This is a more elaborate version of the Ark at Cape Cod, and was built by a team from the New Alchemy Institute. It is essentially an experiment in doing more with less: it combines a household into one solar and wind-powered structure, a greenhouse as part of the main dwelling, a micro-farm including production aquaculture and a vegetable greenhouse and tree nursery components, a waste purifying system and a research laboratory.

In some important respects the Ark, as a dwelling, is the antithesis of the conventional dwelling. We mostly live in houses heated by fossil fuels; the Ark uses the wind and the sun. We waste a great deal of heat; the Ark stores it; we eat imported, and often packaged and processed food; the Ark produces fresh vegetables the year round. (It is also a fish farm.) We produce wastes that have to be treated at high cost, but the Ark turns waste into fertilizer *in situ*.

The director of the Ark project, Ken MacKay, has described some of the practical applications that this experiment is already making possible:

> The problem with greenhouse production in Canada has always been energy costs. In the past, growers have relied on energy-inefficient greenhouses which are heated with fossil fuels. Heating bills alone can exceed $1.00 per square foot annually. This has meant that very few food crops can be raised profitably in Canadian greenhouses. In fact, it now probably takes 10 times as much energy to produce a head of lettuce in a conventional greenhouse in Ontario as to grow the same head of lettuce in California and ship it here. Therefore design and management of solar greenhouses is one of the Ark project's highest research priorities.
>
> Our larger greenhouse has been operating for two winters; it maintains temperatures well above freezing, even on nights when the outside temperature drops to $-20°$ or $-25°C$. The only

energy which must be expended is electricity for a circulating
fan, and no supplemental heating has been required.

. . . Incorporation of the aquaculture facility is equally
important to the success of our greenhouse design. In fact,
preliminary mathematical modelling indicates that over 60 per
cent of the winter heat storage comes from warmth stored in the
large, translucent fish tanks.

The greenhouse solar design has enabled us to produce
winter crops of lettuce, broccoli, chard and other cool-weather
plants – heating only with the sun. This year we harvested
tomatoes in spring and early summer with yields roughly
equivalent to those in conventionally-heated greenhouses.[9]

There will be many more spin-offs from the Ark in the years
ahead: on waste management and composting, biological pest
control, fish farming, and building design. Already some of the
Ark's design features have been incorporated into a series of
'conserver homes' on P.E.I., which are cost-competitive with new
low-cost housing on the island but use little more than half the
energy required for space heating.

The work programme of the Institute of Man and Resources is
funded by a combination of provincial and federal groups, and
donations from private companies and foundations. In 1978 its
budget was about $1 million; a small price to pay for what is certainly
one of the most carefully planned and well structured efforts at
energy and food self-sufficiency in existence anywhere in the
Western world.

The other major Canadian initiative in striving for greater self-
sufficiency and self-reliance is to be found in Sudbury, Ontario.
This is a remarkable multi-partite, self-help endeavour which has
taken on the name of Sudbury 2001 and the challenge of making
that region a self-sustaining metropolis founded upon a diversified
economic structure by the turn of the century. Sudbury 2001 is
dedicated to initiating economic development through appropriate
technology.

Sudbury is a community of 170,000 people in the city and its
neighbouring municipalities. It has long been the centre of nickel
and other mining operations and its largest single employer is
Inco, the International Nickel Company. Along with Falconbridge
Nickel Mines, the mining industry constitutes approximately 25

per cent of the labour force in Sudbury. Contrary to popular belief, Sudbury is not strictly a 'mining town', as can be seen from its survival through a nine-month strike in 1978–9 at Inco. However, this does not preclude the need for more diversification.

Sudbury 2001 began to take shape when a small group of local citizens decided that the only avenue open to the community was to take charge of its own economic development. That conviction began to germinate in the course of preparing and presenting to the Province of Ontario an official response by the Sudbury and District Chamber of Commerce to the province's proposed economic development plan for north-eastern Ontario. The plan was unveiled in 1976; the Sudbury Chamber of Commerce brief was dated April 18th, 1977; the brief bears the title of 'A Profile in Failure'.

The brief pulled no punches in its condemnation of the provincial plan. It referred to the province's proposal as 'a troika of nos – no strategy, no analysis, no programme. Therefore, no use.'

While opposed to the provincial proposal, the brief concluded with a positive programme designed to establish a framework for the development of north-eastern Ontario. It cited some dozen and a half points of reference for such a programme. In so doing, negativism and reaction shifted to 'positive action'.

There were four key figures involved in this movement: one from the Sudbury and District Chamber of Commerce; an economist with the regional government; the publisher of a weekly newspaper; and a senior trade union leader in the region. Although Sudbury had been aware of the need to diversify its economy for years, the report marked a concrete step forward. The second step was taken in the autumn of the same year when the authors of 'A Profile in Failure' expanded their multi-partite coalition to include more community leaders from business, labour, government, academia and other key interest groups, and became known as the Sudbury Committee. What brought them together was a combination of their dedication to living in Sudbury and the realization that continued dependence on the nickel industry, subject to the fluctuations of global supply and demand, would mean the ruin of the economy. This realization was intensified when Inco laid off nearly 2,000 employees on October 20th, 1977.

At the same time, the Sudbury Committee set in motion plans to hold a spring conference on economic development so that the community would have an opportunity to voice its opinions on the direction of the future economy of Sudbury. The success of the April 1978 conference was overwhelming, as over 1,100 Sudburians participated in discussions and workshops which demonstrated their dedication to working towards the improvement of the quality of life in the community.

At the end of the conference, the provincial government pledged $600,000 over the next three years to support technical research, development and demonstration and the continued self-help initiatives.

The Sudbury Committee became the Executive Council of Sudbury 2001, whose task is to oversee the activities of the three major divisions of that organization: the Self-Help Division, the Research Division and the Community Development Fund.

The main function of the Self-Help Division is to encourage community participation in economic development by enlisting volunteers to implement selected programmes on their own initiative. Sudbury 2001 is not an organization in the traditional sense which is going to do things for the people. Therefore, the success of the endeavour depends on the community's initiative to take charge of and utilize the 2001 infrastructure to achieve their objectives.

After the conference, various task forces were formed on marketing, culture and the arts, and image improvement. The substantial volunteer efforts were later supplemented by 'Canada Works' and 'Young Canada Works', Federal funding secured to employ students and adults to assist in the implementation of the task force's ideas.

The Image Improvement Task Force has been responsible for most of the activity in the efforts to improve Sudbury's environment and eliminate the 'smoke and black rock' image the media have chosen to paint of the city.

Guided bus tours of the city were organized and operated during the summer of 1978 and 1979, as were demonstration tree and flower planting projects. The 'History of Creative Business and Enterprise in Sudbury from 1883–1979' was documented for use in elementary history programmes so that schoolchildren may learn of their community's past accomplishments, and entrepreneurs and draw from their example. Graphic design students were employed to design promotional materials for tourist groups, and a slide show of the area was assembled.

The Marketing Task Force supervised a group of students during the summer of 1978 and 1979 in implementing a 'Buy Local' campaign designed to promote awareness of locally produced and manufactured goods and stress the significance of keeping Sudbury-earned dollars in Sudbury. Two 'Meet the Makers' shows provided local manufacturers and craftsmen with an opportunity to display their products, and an Industrial Directory was compiled.

The Culture and the Arts Task Force, a group which would have gladdened the heart of Schumacher, who argued that without a strong local culture, nothing else can grow, began working on establishing a folk coffee house and an arts council. These tasks have since been absorbed by other existing cultural organizations in the city, reflecting the way in which the community can utilize the 2001 infrastructure to achieve their own objectives.

During the summer of 1979, a group of students was employed in the recycling of newspaper. Depots were set up through a network of playgrounds, service clubs and churches where people could drop off their newsprint. It was later collected by a local business to be used to make insulation. Another group of students took on the task of gathering and disseminating technical and economic data on the feasibility of solar energy in the Sudbury area.

In order to broaden the base of community support for 2001 and increase citizen participation, a Neighbourhood Development Task Force was formed in March 1979. The region was divided into various neighbourhoods and municipalities and 'town meetings' were held at which neighbours were given the opportunity to reflect on past accomplishments, present challenges and design a plan for social, cultural and economic improvement in their particular area.

Two other projects of relatively short-term activity but long-term impact are the compilation of an economic atlas of the Sudbury Region and the establishment of a tourist park at the entrance to the city.

With the assistance of the Science Council of Canada, Sudbury 2001 has set up an international network of over 700 analysts in the various fields of appropriate technology and alternative development.

Long-range plans for Sudbury 2001 centre around the two other major divisions, the Research Division and the Community Development Fund. In order to finance the activities of 2001 beyond the first three years of the provincial funding, a Community Development Fund is to be established by raising private and public capital. Besides financing the internal activities of 2001, the fund will also assist in the funding of industries and small businesses that are unable to secure adequate financing, by providing last-resort equity and/or loan financing and/or loan insurance.

In the Research, Development and Demonstration Division, an international search for a multi-disciplinary team of analysts to undertake quantitative and qualitative analysis of the structure of the economy and recommend feasible economic development projects is under way. Not content to wait until this team is fully operational, Sudbury 2001 launched two major projects within the first eighteen months of its existence.

The first is a mohair industry which will entail the raising of angora goats, processing their hair into mohair wool, and the eventual manufacture of textiles. This project is now undergoing a feasibility study in such areas as financing, technical requirements and marketing. Preliminary examination of the project indicates significant potential in terms of job creation for farmers in the area. It is expected that income from the raising of goats will provide the incentive to return to farming on a full-time basis. If this is successful it will demonstrate the establishment of a self-contained localized industry – from raw material to finished product.

The second project, also in the feasibility study stages, will be a 'Quadractor' manufacturing plant. The Quadractor, invented by Canadian Bill Spence, is a lightweight machine, weighing only 500 lb. and yet able to move a 4,000 lb. payload. Its features include its low cost ($3,000), low consumption of raw materials and fuel (it can run for eight hours on two and a half gallons of petrol), versatility (ploughing fields, hauling logs, ploughing snow, etc.), and the simplicity of manufacture, operation and maintenance. This all-round agricultural machine, now being manufactured only in Vermont, U.S.A., is truly the 'mechanical mule' of appropriate technology. Sudbury 2001 plans to finance its half of the project by raising funds within the community by the sale of shares.

At first sight Prince Edward Island and Sudbury have little in common. But, allowing for the fact that the Sudbury experiment is only just beginning, while the P.E.I. programme is well under way, they have a good deal more in common than meets the eye. Both have been told, as it were, by the large-scale technology regime that they can really play no role except as recipients of welfare payments of one kind or another. But neither community – they are both really islands although only one is surrounded by water – will accept that they are pre-ordained, or perhaps the word is programmed, to become second-class citizens. They have decided instead to show that small is possible, and the future is on their side.

Part Three

WHAT SMALL MAKES POSSIBLE

7

ECONOMICS AS IF
PEOPLE MATTERED

Both in the rich and the poor countries, the idea of intermediate or appropriate technology is now entering the consciousness of economists, administrators and politicians. There would seem to be four stages in the process. It starts with the widespread rejection of the concept, because it means a radical break with conventional behaviour. The second stage is general acceptance of the idea, but little support from government, or international institutions. The third stage would be active involvement on a considerable scale to mobilize knowledge of technological choices and to test them under operating conditions; and the fourth would be the application of this knowledge on a scale that makes it not exceptional or 'alternative' but a normal part of administrative, business and community activity.

As far as developing countries are concerned the idea is now in its second stage and hovering on the edge of the third. Most poor countries are now aware of the extent to which they have become economically dependent upon the industrialized countries, and most see the need, if not to shake themselves completely free, at least to become much more self-reliant and regain their integrity. But as the recent United Nations Conference on Science and Technology has revealed, most governments of developing countries are still far too preoccupied with questions of international trade, and with securing the technology of the multinational companies as easily and as fast as possible. This is a sort of tunnel vision that leads to aid and development policies which, in the event, bypass the rural areas of developing countries and thus bypass the very source and centre of their poverty. To say this does not mean that international trade is irrelevant to the problems of development. But a considerable change of emphasis is still needed. As long as some 80 per cent of the population of most developing countries remains caught in the vice of rural poverty and stagnation, no amount of stimulation of international trade, nor even a significant change in the terms of that trade, can have a real impact on the problems of world poverty.

The very poor can indeed be helped to help themselves, but only by a deliberate and systematic effort to get into their hands the tools and equipment with which they can furnish themselves with food, clothing, shelter and other basic necessities of life. What people do for themselves and for each other, as Schumacher reminded us in *Small is Beautiful*, will always be more important than what they do for foreigners.

There is one sector of the international community that has fully understood, and is acting upon, the overriding need to provide the poor with the means of working themselves out of poverty, and that is the community of the voluntary, non-government organizations. Anyone who doubts this should study the non-government organizations' report to the U.N. Conference on Science and Technology held in Vienna in August 1979. The bulk of its recommendations comprise practical suggestions to promote self-help among the poor, and self-reliance for developing countries along that route.

Nothing much can happen by way of the widespread introduction of appropriate technology in any developing country until it starts to build up its own capacity for autonomous technology choice. This does not mean creating a string of government research establishments, which are singularly ineffective in helping the poor. It means creating appropriate technology units such as those described earlier, in India, Africa and elsewhere, that are capable of doing the four things that are essential for the diffusion of the techniques of self-help.

The first is the identification of needs, as perceived not by foreign experts but by local people themselves. The second is the mobilization of the knowledge required to meet those needs. Ideally, such knowledge should be available within the country itself. If not, the A.T. unit must know where to get it, backed up, if need be, with the technical expertise needed to apply it. Next, once a successful pilot scheme has been set up, there is the task of extension – of making the innovation widely known, and of being able to assist other individuals or groups to take it up. Fourthly there is the creation of supporting facilities, primarily those of a continuing service of advice and improvement, and of ensuring that necessary manufacturing and repair facilities exist within easy reach. Every successful A.T. unit combines these four functions in one form or another, irrespective of how simple or complex the technology concerned.

It is the universal experience of all who work in this field that the most effective A.T. units everywhere are bodies that have the maximum possible freedom both in getting information and in how they apply it. These qualities generally exclude government-run organizations, which simply do not have the necessary flexibility, speed of action, and the ability to be selective and non-authoritarian. The reader need only look at some of the most effective A.T. units in existence – in Ghana, India, Tanzania and Kenya, for example – to verify this. Although there may be considerable government support for the work being done, the A.T. units have been kept free of bureaucratic procedures or control.

Appropriate technology units of this kind are the keystone of any policy of rural development worth the name. In large countries such as India there should be at least one in each State; and each of these should have its links with local institutions of teaching and research, local industries, voluntary organizations, official research establishments and overseas A.T. units. That is, it should be able to draw upon the facilities of government, industry, and other sources of technical knowledge without restraint.

The first step, therefore, in a thoroughgoing policy of bringing industry into the rural areas would be that of fortifying and making more complete a network of A.T. institutions which is already partly in existence, and whose experience is there to be drawn upon.

A second part of such a policy would be that of creating the environment in which appropriate technology can be most effective. Almost without exception the infrastructure in developing countries, both the physical infrastructure in the form of, say, roads and provision for health and education, and the administration and fiscal and financial procedures, have been modelled on those of the industrialized countries. That is to say, orthodox, conventional economics, technology and applied science, and administration, are designed to serve an efficient system of production and consumption and *not* to develop the capability of people to look after themselves. The conventional structures are those that support a system of economics whose starting point is the production of goods, and for which purpose labour is a factor of production.

A structure that supported an economic system whose starting point was people – economics as if people mattered – and regarded goods as the natural and inevitable result of making everyone productive, would look very different.

Thus if people and not goods were made the point of departure, very high priority would be given to improving the technologies available to the women of developing countries, who do much of the most arduous work, but who are the forgotten majority as far as appropriate technology is concerned. Much more serious attention would be given to work-based education – what is being done in Kenya and Nigeria are shining exceptions so far – and a major effort would go into both improving the technology of small farming and peasant holdings, and into low-cost ways of helping communities, and each country, to become self-sufficient in food. For human and environmental (as well as cost) reasons, this essential attention to the needs of the small farmer would be based on minimizing the use of inorganic fertilizers, herbicides, pesticides and other lethal products of the chemical industry, and instead bringing more science to bear on developing biological husbandry. Likewise, local processing of food would be given very high priority, to the benefit of local incomes, employment and nutrition. Local roads, and better vehicles to travel on them, would be given precedence over trunk routes and long-distance transport. The same considerations would apply to services such as health care, water supply, and credit and marketing facilities, and to energy supplies and small industries.

In all these areas of human activity, and more, it has been shown that there is already a record of solid achievement. A people-oriented technology is no longer a pious hope: it is a practical possibility, and its widespread adoption could now be supported by governments and the international community on a really large scale.

A decentralized economy which put production by the masses

before mass production would also require changes in the administrative structure. Developing countries which supply the industrialized world with much of its raw materials are today clearly aware that their best interests are not served by continuing to export raw materials in their natural state: that they should progressively secure the benefits of value added, by processing the raw materials before exporting them. But it is as yet less widely accepted that the same holds good *within* a country. The rural areas will remain the poor relations of the cities, and life in them will continue to deteriorate, unless they get new work and new income opportunities. But everywhere the administrative structure, the rules of the game, so to speak, favour the large against the small, the centralized operation against the decentralized, the rich against the poor. They bear the imprint of a system of technology and economics that puts goods before people. These rules relate to such matters as the availability and terms of finance, access to research and development facilities, management and technical training; freight rates and transport facilities; taxation policy; foreign exchange regulations; and economic planning, especially the criteria employed in assigning priorities to different kinds of economic activity.

A simple example of changing the rules is the Indian government's decision to reserve a list of specified industries for development only in the small-scale and rural sector. Another (cited by Schumacher in *Good Work*) is the concept of the Chinese commune. This can be described as a 'cultural/industrial administrative unit' (it has parallels in the Tanzanian *ujamaa* village and the Israeli *kibbutz*) and its creation as a relatively self-sufficient geographical area is in itself a major change from the conventional rules. The Chinese have carried this to its logical conclusion by saying, in effect, 'If you want to import anything into the commune, prove first that you can't make it (or something that will do the job) for yourselves.' In contrast, the conventional 'goods-oriented' economics says, 'If you are in the rural area and you want to make something, first prove that it can compete with city-based production.' Is it surprising that the Chinese have full employment – albeit at a low level of individual productivity but with everyone producing something, while most developing countries, locked into Western notions of economics, stand helpless in the face of mounting unemployment and rural decay?

Developing countries could help themselves to change the 'rules of the game' by setting up their own technology assessment units. Their task would be to advise their governments about the choices of technology available, and about the implications of different choices. Thus a developing country intending to introduce cement manufacture, say, or sugar refining, would be advised on the choices open to it – in these instances, one huge factory or forty or fifty small units. What would these options imply in terms of availability, foreign exchange costs, running costs, social impact, employment, local development and income distribution, and so on? Ultimately it is only when developing countries start to demand that real choices of technology should be offered to them, and to assert this as their right, that they can begin to break away from their dependence on industrialized countries: they would then be free to choose technologies that maximised their self-reliance.

If the developing countries are well into the second stage of our 'four stages of appropriate technology', the industrialized countries are still at stage one. With a few exceptions in Britain, and a few more in the U.S.A., the official attitude towards appropriate technology and the alternatives groups that are applying it is one of rejection or antagonism.

It is true, of course, that for highly industrialized countries a changeover to technologies that are smaller, more decentralized, more humane and less demanding on non-renewable resources represents a major departure from orthodox thinking and conventional practice. The developing countries, after all, still have many of their options open. The rich countries are well along the road towards an economic and industrial system that looks less and less sustainable.

One of the principal characteristics of industrialization, as most people have experienced it, is its overriding tendency to create a more and more *dependent* population. This dependence, this external direction of people's lives, is most evident in the case of their employment, over which they have virtually no control, either as regards its availability or its quality. It also applies to other aspects of daily life. There are always two systems by which we support ourselves – the 'self-care' system and the market system; the latter requires us to earn money, in order to buy goods and services produced by others. The self-care system has declined to near-vanishing point, and the result is a great deal of waste and expense, and a loss of independence for the family and the community. The same applies to education, health care, and recreation.

All this is very discouraging. But its positive aspect is that it has caused a large and growing number of people in Britain and in North America, as we have seen, to start reversing this trend by launching experiments of all kinds in the direction of greater individual and community self-reliance. This has produced what can only be described as a flowering of creative activity on the part of tens of thousands of people.

The origins of these groups working on alternative technologies and life-support systems are varied, and their attitudes to conventional institutions and ways of doing things are by no means invariably sympathetic. Some have come together out of a growing concern for the environment; others have their primary focus on energy conservation, housing, health, agriculture, local manufacture. There are now few branches of human activity where the conventional mode of doing things does not have its counterpart within the alternatives movement. The variety of their origins and methods of working should never blind us to the fact of what they have in common, namely, a recognition that conventional industrial society is on a collision course with human nature, with the living environment, and with the world's stock of non-renewable resources.

So long as there is no general acknowledgment that we *are* on this collision course, this upsurge of groups working on alternatives of all kinds will tend to be written off, by government and other power groups, as at best a way of life for eccentrics, at worst a serious threat to economic order and discipline.

It would be idle to pretend that the transition from the 'limitless growth' economy of the recent past, or the 'stagflation' economy of the present, to the conserver society of the future will be plain sailing. But it would be equally misleading to argue that policies favouring the widespread adoption of appropriate technologies would imply some drastic and unacceptable collapse of living standards and life as we know it.

Consider, for example, a few of the steps that Britain, or any other highly industrialized country for that matter, could well take now, that would in practice set us on the way to a more sustainable future.

One would be to start a major programme of energy conservation, with initial emphasis on reducing the 40 per cent or so of primary energy production that now goes into space heating; and at the same time to further the development of renewable energy resources on a really significant scale. (In Britain, this would enable us to avoid incurring the enormous cost and unimaginable dangers of nuclear power.)

A second part of such a policy would be to start a broadly based programme of research and experiment aimed at liberating agriculture and food production from its present very substantial dependence on fossil energy, and making Britain self-sufficient in food to the maximum possible extent.

A third component would be to promote and facilitate the localization and decentralization of manufacturing and service activity, both as a means of creating local employment, and of cutting down the escalating costs of long transport hauls and over-centralized services. It is to promote action along such lines that the Intermediate Technology Group celebrated its fifteenth anniversary by launching, early in 1980, a campaign to start the Schumacher Centre for Technology Choice.

Could really effective action along these lines do anything but enhance the quality of life?

It will also be evident that if such deliberately decentralist and conserver policies were to be followed, then appropriate technologies, and most of the current activities of the alternative groups described in this book, would fall into place as perfectly obvious ways of attaining these objectives.

In fact this relatively unknown part of the economy represents an important part of present-day reality. It is opening up the way to what James Robertson calls the sane, human and ecological society, which recognizes and encourages the variety of human resources and abilities that exists in all communities.[1] In contrast, the mechanistic 'single-solution' approach of conventional economics – modelled on the physical sciences – typified by the centralized manipulation of aggregate demand, and by the concept of gross national product, is by its very nature incapable of recognizing this variety or of freeing the creative energies of large numbers of people. No amount of juggling with the monetary, fiscal or price system can provide useful and satisfying work for the people who are losing their jobs through the increasing capital-intensity of large-scale industry, or through growing shortages of resources; nor can it provide any answer to the alienation of the workforce, to inflation, or to the degradation of the environment.

Can we not recognize that there is really no other choice than to create a new technology and economic system designed to serve not a continuously escalating spiral of production and consumption, but to serve people by enabling them to become more productive? This is precisely what is being attempted by such groups as the Local Enterprise Trusts in Britain, for example, and by the Lucas Combine, by small co-operatives and common ownership firms on both sides of the Atlantic, by Sudbury 2001, by, in short, the alternatives movement. The rich countries need more of this kind of work at least as urgently as the developing countries.

Supplement A

APPROPRIATE TECHNOLOGY ORGANIZATIONS IN DEVELOPING COUNTRIES

The choice of technology is the most important collective decision confronting any country. It is a choice that determines who works and who does not; that is, who gets income, new skills, self-reliance. It determines where work is done, whether concentrated in cities or more decentralized in smaller units; that is, it determines the kind of infrastructure required, and the whole quality of people's lives. It determines the ownership structure of industry; huge technologies are available only to the rich and powerful, whereas small technologies are tools in the hands of the poor. Technology, in short, carries its own culture with it – it is certainly neither economically nor culturally neutral. For a developing country, generally speaking, the larger and more complex the technologies it gets, the greater becomes its dependence, economic and cultural, upon the rich industrialized countries.

The ability to exercise choice in the matter of technology is therefore of the highest importance to any developing country. Choice of technology is, however, precisely what is denied to the developing world by conventional patterns of aid and development. That is why the growth of appropriate technology centres in the rich countries must be matched by similar centres in the developing countries. Between them they can open up technological choices where none existed before.

The work of some of the pioneering appropriate technology groups in the developing countries is described below.

The Indian Sub-Continent

India

It was in India, as we have seen earlier, that Schumacher first developed his concept of an intermediate technology. But at that time, in 1963, the supporters of rural industrialization through intermediate technology were very much in the minority. The economic planners (then in the ascendant), the Planning Commission, and the government in general, were against the idea. A brave effort was made by the Rural Industries Committee of the Planning Commission to introduce intermediate technologies through the forty or so special Rural Industries Projects that were then being started throughout the country; but lacking any official support from government departments, official research and development organizations, or universities, the officers in charge of the projects could do little. There was no reservoir of experimental work, far less any body of practical, tested information on which they could draw. For nearly the next fifteen years the supporters of intermediate – or as it became known in India first, and later elsewhere, appropriate – technology fought a losing battle for official support. Whatever work was done to develop and introduce small-scale, capital-saving industries that would fit the needs of rural communities, was done, so to speak, against the mainstream of economic policy, chiefly by the Khadi and Village Industries Commission, the Planning, Research and Action Institute, Lucknow, and other mainly voluntary organizations, and dedicated individuals.

It is only quite recently that this climate has changed. Rural development, and employment through the use of appropriate technologies, is now a central part of official policies and programmes.

It is now widely acknowledged, then, that India's massive and growing army of unemployed and under-employed people demands the creation of millions of new and more productive workplaces. In practice, this means making available to rural people better tools, equipment and facilities, on a very large scale. But what are these technologies that will in fact raise the incomes of the rural poor, on a sustainable basis? Who can produce these technologies? How can poor people get access to them, and own them? These are the questions to which a growing number of appropriate technology groups in India are addressing themselves.

Several teaching and research institutions in different parts of India have started A.T. programmes. Two relative newcomers to the field – both started in 1974 – are the Indian Institute of Technology, Bombay, and the Indian Institute of Science, Bangalore.[1] They are significant not least because both are science and technology centres of high repute, and until recent years neither would have considered rural technology to be worthy of their attention.

The Appropriate Technology unit in the Indian Institute of Technology, Bombay, was started by a young physicist, Anil Date, after students and faculty members had worked in the rural areas during the drought of 1972–3. Its chief activity is running undergraduate research projects on which final year students can spend six hours a week.

During the past four years about forty students and fifteen faculty members have designed, built and tested more than twenty items of low-cost machinery and equipment. These include a small (¼ horsepower) hot air solar engine, and a diesel engine adapted to use 80 per cent methane. Their work on building materials has established optimum standard mixes of materials used in making fly ash building blocks, lime-pozzolana mortars, and paddy-husk ash cement. Two of the items they have designed for the very small farmer are a paddy-drier and a mini-plant for citronella oil extraction. The paddy-drier handles about 200 lb. of paddy in three hours and is especially suited to small farmers growing early varieties of paddy in the Bombay area. The mini-plant for citronella oil distillation has been scaled down for use by a two-acre farm owner; previously the smallest plant available required fifty acres of grass to make it economic.

Whereas all the work of the A.T. unit in Bombay is done on a voluntary basis, the research and development programme of ASTRA (Application of Science and Technology to Rural Areas) at the Indian Institute of Science, Bangalore, is more closely integrated with the parent organization. ASTRA was started by a distinguished scientist, Professor Amulya Reddy. Its activities include projects sponsored and funded by external organizations (such as the Tata Energy Research Institute): research on biogas, windmills, bullock carts, rural energy use patterns; bamboo conservation, energy planning, hand pumps and village ecosystems, all of which involves seventeen members of the faculty. Others are engaged on in-house research on alternative energy sources, low-cost building, small-scale soap production, educational aids for science teaching. Then there are student projects which form part of the Master of Engineering course, including investigations into biogas production, the extraction of sodium silicate from rice husks, plastics from castor oil, cellulose fibre from groundnut shells, and various aspects of low-cost construction materials and methods.

ASTRA is also erecting workshops within a rural extension centre at Ungra, a village some eighty miles from Bangalore; this centre includes housing for five permanent members of the staff from ASTRA, and for others on short-term assignments.

So far, ASTRA's work on hand pumps, which identified causes of hand pump failure, has resulted in their design being adopted by Karnataka State for all future installations.

These institutions, the I.I.T.s and ASTRA, recognize that there is a powerful tendency in academic circles to identify and work on problems that are of interest to scientists, rather than to start by working with people, identifying their problems and needs, and evolving solutions that are not only technically but also economically and socially acceptable to rural communities. But although neither of them has been going long enough to show what they are really capable of producing, both have already done something of great value by demonstrating that work on small-scale rural technologies can be every bit as exacting and scientific as work on conventional technology.

Other scientific and technical institutions have been doing the same thing in different ways. The Birla Institute of Technology in Ranchi, Bihar, has been building up a very effective little A.T. unit, SIRDTO (Small Industries Research Development and Training Organization) since the early 1970s. This started as a programme for training young graduates as entrepreneurs, and setting them up in business on the campus. It has since expanded into work on rural projects, agro-based industries, and the training of rural artisans.

Under the graduate scheme, thirty small factories have been started, twenty-six on the campus and four in the locality near by.

Ranchi is the centre of the coal and ancillary industry in India, and many of the big firms use imported equipment. The Birla graduates do projects in their final year which are related to those industries' needs. Each project aims to identify and solve a technical problem of a large company, and to make better use of indigenous resources – often by import-saving. Once the prototype has been field tested and accepted by the user, the young entrepreneur is helped to go into production on a commercial scale in one of the campus workshops. These are built by one of the young entrepreneurs who started a brick-making and contracting business.

Generally they make specialized products or components which local big industries need but cannot produce for themselves, such as specialized castings; mica-insulated and heatproof tape; flameproof electrical switchgear; steel conveyors and safety equipment for use in coal mines, and so on.

Having started in this way, many of the small units have widened their markets and their ranges of products. A small service centre has been set up by the Birla Institute near by to provide facilities for product development, design and testing, including producing printed electrical circuits to order.

The setting up of these small firms led to a need for training local operatives. At first this was done solely for the firms but later a training programme was started on simple metal working, and bicycle and diesel pump maintenance, to enable villagers to meet the growing demand for such skills. This in turn has led SIRDTO to explore ways of making better and cheaper tools for local artisans.

So far all the small units have been engaged in engineering of one kind or another; now one is starting up that will process local raw materials. On a ten-acre plot they are growing medicinal herbs, and grasses and seeds from which essential oils can be distilled, such as citronella and lemon grass. A young graduate is starting a small unit which will draw its raw material from villages in the locality. The idea is to provide villagers with small distilling units, so that they can do first-stage processing of the grasses and thus add value to their product and increase their incomes. More refined distillation and preparation of products for the market will be done by the small firm located on the campus.

Often quite simple innovations can have big effects on village incomes. Villagers' earnings from the collection of lac (an insect residue from certain trees) have been more than quadrupled by a very simple cleansing process suggested by the university staff. It may also prove possible to make paper from the grass residues after distillation.

The university staff are also giving systematic attention to the growing and processing of tussor silk (wild silk, the cocoons being cultivated on local trees). For this another plot has been started on the campus, improved processing machinery has been developed, and villagers are being taught how to spin the thread into yarn at a small workshop adjacent to the plantation.

The SIRDTO team works closely with the Xavier Institute which has long been established in the area. The Institute runs a food-for-work project aimed at raising food production, and especially vegetables for villagers' own consumption and for sale, by means of small-scale well irrigation. To assist this programme, the SIRDTO team designed a simple method of building small dams, using local stone and incorporating arches in the construction to give the dams maximum strength. Many such small dams, some only a few feet long, built across gulleys and streams, will arrest the rapid run-off of rainwater that now takes place, and raise the water table. This will make the well-digging programme much more effective.

The SIRDTO unit is run by a small group of senior faculty members from the Birla Institute, headed by the Institute's Director, Dr H. C. Pande. Faculty members are encouraged to take part in all programmes.

Another teaching institution which is deeply committed to rural development in a practical way is the Allahabad Polytechnic which, with 4,000 students, is probably the biggest polytechnic in India.

The Polytechnic has a thriving production and training centre staffed by some 400 technicians and skilled workers, through which students are introduced to practical work. The centre has an annual turnover of about £150,000. It makes, assembles and markets a wide range of products including cheap and robust sewing machines, transistor radios, agricultural tools and equipment, 'knock-down' furniture.

The Polytechnic's principal, R. N. Kapoor, is convinced that every technical institution in India should be helping to solve the problems of rural communities. Under his guidance Allahabad Polytechnic runs an integrated rural development programme covering fifteen villages (about 1,000 families) which lie in several clusters around Allahabad. Their object is to find ways of helping the villagers to become as self-sufficient as possible, to develop a methodology that can be applied by others.

There are Polytechnic staff living in each village cluster. Staff and students work with the villagers on local road improvements, sanitation and water supply projects and health and vocational training schemes. A visit to a typical village cluster reveals a variety of modest but practical improvements: an agricultural tools workshop and blacksmithy; a village cobbler equipped with better tools, two handlooms for production and training at village level; a service centre (one of six serving the fifteen villages) which demonstrates and lends out improved farm tools, runs a garment-making class for women and a weekly market, and provides for visits by a doctor and health worker. On waste land donated by the village a clinic is being built with local materials, lime being made on the site.

At the Polytechnic there is a training workshop where young people from the villages can learn to make carpets, soap, chalk; and some of them are attached to the production and training centre to learn more advanced skills. There is also a sheltered workshop for the handicapped.

To support this rural development programme, a special A.T. unit was started early in 1978. This makes and develops equipment such as a solar rice cooker which can cook 2 lb. of rice in forty minutes, a red clayware water filter, and a metal winnowing fan; they are testing a sisal rope-making machine for the Khadi and Village Industries Commission, and their building unit has developed a variety of low-cost building components – door frames, roofing slabs and joists – using ferro-cement.

The four institutions whose involvement with appropriate technology has been described above – at Bombay, Bangalore, Ranchi and Allahabad – are not of course the only centres of teaching and research where work on similar lines has been started. But it should not be imagined for a moment that they are representative of what is going on in universities and technical institutes in India generally. What they do represent, in their different ways, are determined efforts, by a handful of exceptional people, to show that the best minds in the country can and should be productively engaged on working not for an élite but for and with the mass of the people in the rural areas. They are proceeding, by trial and error, towards a higher education, not for privilege but for service.

THE APPROPRIATE TECHNOLOGY DEVELOPMENT ASSOCIATION (ATDA)
The organization which has done more than any other to promote
the concept and practice of appropriate technology in India, not
only within the universities and technical colleges, but also in
government and industry, is the Appropriate Technology Development Association, Lucknow. Its origins go back to 1967, when an
Appropriate Technology Development Group was formed in
Delhi by people from industry, research organizations and government service. That was disbanded after a few years owing to the
absence of internal or international support, but it led to the
formation of two other A.T. bodies. One was the Appropriate
Technology Cell in the Ministry of Industry, which for several
years was merely a token presence but continued to keep the idea
alive in government circles, and now funds several A.T. programmes.

The other was an independent group, formed on the initiative of
the late J. P. Narayan.[2] This was the Appropriate Technology
Development Association, which started life under the wing of the
Gandhian Institute of Studies in Varanasi in 1972, and two years
later moved to its present location in Lucknow.

The Association's objectives and methods of work express the
ideas of two remarkable Indian engineers, M. K. Garg and
Mansur Hoda. Their starting point is the need to raise incomes in
rural areas by creating new and more productive employment.
More precisely, it is to find ways of doing this on a sustainable,
self-generating basis by bringing industry into the rural areas, of a
kind that can be owned and run by people in the rural communities.

For such industries to become widespread, they argue, their
products must be of a quality that is generally acceptable; and their
technologies should be as small and simple and capital-saving as
the best engineering skills can make them.

If they are to be competitive, and to multiply without continuous subsidies, rural industries must be furnished with the same
sort of back-up facilities that are currently available only to large-
scale industries: with suppliers of machinery and equipment, after-
sales service, consultancy and training and research services. The
Association's aim is to produce technology 'packages' of this kind
for rural industry.

All this must seem very ambitious. But there is nothing theoretical about it, because such technology packages have already been made available to two industries – sugar and pottery – with great success, and in both instances are opening up the way for continuous improvement and development.

The most striking example is the development of mini-sugar plants. A pilot project for a mini-sugar plant making crystal sugar (which could not be made by the traditional process) was started in 1957 by M. K. Garg, who was then employed by the Planning Research and Action Institute, Lucknow; after several years, during which the bulk of the development costs were borne by a number of private and co-operative small-scale sugar units, an integrated mini crystal sugar plant was developed. The new technology spread rapidly. Today there are some 5,000 mini-sugar mills in India, producing 1·3 million tons of sugar a year – about 20 per cent of India's total output – and employing more than half a million people. This employment is at its maximum during the winter, the slack agricultural season; and the mini-plants are located in remote rural areas where the problems of unemployment and under-employment are most serious.

A number of well established machinery manufacturers have taken up the manufacture of the mini-plant equipment and can offer a good service of technical assistance for setting up the plants. Consultancy organizations, both government and non-government, can provide a turn-key service as well as operational advice.

The social and economic contribution of the mini-sugar technology may be summarized as follows:

For the capital costs of one large sugar mill, forty mini-plants can be built;

the forty mini-plants produce two and a half times more sugar than the big mill;

they employ ten times more people than the big mill.

In terms of capital output and employment generation, then, the mini-plant is unquestionably superior. But that is by no means the end of the story. The social effects are probably even more important, though they are generally ignored by economists because they cannot be accurately measured or quantified. An

I.T.D.G. team that recently visited a new mini-sugar plant in South India, near Hyderabad, found that:

> the mini-plant fills what was formerly a gap in the rural economy. In this area, the paddy season runs from June to October. Between November and May there is virtually no work, so many of the younger men moved off to the cities looking for jobs, some never to return. But things have changed since the mini-sugar plant was started. It operates from November to May, providing about 160 new jobs during those months. This alone brings at least Rs 150,000 new income to the villagers, as well as sustaining the social life of the community.[3]

Although this industry is now well established, it still lacks the kind of research and development support that is taken for granted in large-scale industry. This is a deficiency which the ATDA is now making good with the support of the State government, as will be described later.

The other example of successful decentralized production is the manufacture of whiteware pottery. In this instance a 'service centre' approach was adopted, in order to provide village potters with raw materials, and to develop adequate facilities for firing the whiteware.

The raw material of whiteware pottery is not the clay, which is everywhere available to the village potter, but a finely ground mixture of china clay, quartz and felspar. The preparation of this 'body' needs mechanical processing beyond the reach of an individual potter. This processing plant was the nucleus of the service centre. The first one was set up at Khurja, near Delhi, in the 1940s. The potters collect the prepared 'body' from the centre and shape it in their cottage workshops. The centre also supplies them with improved equipment: mechanized potters' wheels, and jigs and moulds now in common use.

After the wares are shaped they are fired. Here again a central facility was provided, in the form of kilns which could fire to the required temperatures (a minimum of 1,100°C; higher than is possible in a traditional village potters' kiln). The Khurja centre also stocked glaze, and set up product design and marketing facilities. Starting with only a few potters, Khurja has since grown to more than 400 cottage workshops employing 20,000 people.

The Khurja centre was a major step towards scaling down the whiteware industry. But before further scaling down could take place another technological problem had to be overcome. The minimum economic size of the kiln required the output of about ten potters to keep it going, and for reasons of cost and fuel efficiency, kilns were grouped into threes (each group being served by a large brick chimney stack). Thus the minimum cluster was at least thirty potters in the same community; and this was only possible in semi-urban areas. How this barrier was surmounted provides a good illustration of the fact that small-scale technologies will often demand the highest qualities of engineering and technical skill. After nearly six years of experiment, a way was found of reversing the flow of heat within the kiln, carrying the fumes up through a double-wall in the kiln, and providing each kiln with its own small chimney on top. The result is that an eight-foot-diameter family size kiln can now operate as efficiently as the conventional eighteen-foot diameter kiln.

The new kiln designs were in fact worked out at another centre, at Chinhat near Lucknow. This centre showed that smaller, semi-rural clusters of whiteware pottery production could operate successfully, and this pattern has since been adopted in several other centres in India.

In these two instances, it took several years to create small-scale, decentralized industries capable of standing on their own feet. There are other cases, especially among the traditional craft and artisan industries, where it should be possible to get results more quickly. The Association's programme comprises:

scaling down large scale technologies;
scaling up traditional technologies;
developing home living and communities technologies.

The starting point of all ATDA activities is detailed surveys and analytical studies to identify areas in which appropriate technologies are needed and can be developed. These 'state of the art' surveys answer such questions as: what is the background, and present status of the industry? What work has already been done to modify or adapt the technologies in question? What gaps exist, what research and development are needed to fill them?

The second stage is to build a pilot or start a field project under realistic operating conditions, as a commercial unit. The next stop is to make available to potential users, and makers of the relevant equipment, a complete package of designs, specifications and technical literature, along with the technical assistance services needed to get the plant or process going. Finally, there is the creation of a research and development support system, to ensure that there is a continuous technical improvement in the industry.

How this works in practice can be best described by looking at particular examples of ATDA's work. By way of introduction to these case studies, a word should be said about which industries fall into which broad categories: scaling up, scaling down, and home and community living improvements.

Many small industries that formerly flourished in India's rural communities, and were the mainstay of a relatively self-sufficient rural economy, are being destroyed by large, city-centred industries. The technology of these rural industries has remained virtually static, and without further development they will continue to decline. They include handloom weaving, village pottery, tanning and leatherwork, carpentry and blacksmithy, vegetable oil extraction. They are candidates for scaling up, being upgraded to make them more competitive.

In other instances, centralized industries have taken over completely, or their technology and its products have been developed to a point at which they have no counterparts at the rural level. Cement, paper, cotton and wool spinning are such cases, where the technologies can be scaled down and decentralized.

Technologies for improving the physical amenities of rural homes and communities include new and improved facilities for water supply and sanitation, small-scale energy supplies for cooking, lighting and crop processing, and animal husbandry.

In the case studies that follow there are examples from each of these three categories.

The textiles industry comes high in the Association's priorities, and some quite spectacular results can be expected from work that has already been accomplished. Next to agriculture, the textiles industry is the largest employer of labour in India; it was also one of the industries most severely affected by the development of large-scale mechanized technology. The adverse effect on spinning was far greater than that on weaving; by about 1940 village spinning had vanished except for hand spinning kept alive under the Khadi and Village Industries programmes. The decentralized weaving industry still produces nearly half the total cloth in India, and employs about 7 million people (as against 1 million in the large mills).

But the handloom weaver now depends entirely on the yarn produced by the large mills and, being at the wrong end of a long line of middlemen, pays between 20 and 30 per cent more for his yarn than do the large-scale weaving mills (which are integrated with the spinning mills). This has cut down the village weaver's income to near subsistence level.

In the past, most of the attempts to help the handloom industry have concentrated on improving marketing, and by subsidizing the cloth. Attempts to lower the price of yarn by setting up co-operative spinning mills were not successful.

In the 1960s the Khadi and Village Industries Commission brought in a leading textile machinery designer to improve the traditional hand spinning machine, the Ambar Charkha. He developed first a six-spindle hand-operated machine, then a twelve-spindle, pedal-driven machine, and finally a twelve or twenty-four-spindle unit driven by a small electric motor: the last was really a scaled down version of the spindles used in large mills. But for this to work effectively it was necessary to prepare the raw cotton to a much higher standard – in fact to the same standard as used in the large mills, where pre-spinning operations employ a lot of expensive and complex machinery. But no small-scale mechanized pre-spinning equipment existed.

This was the technical gap that the Association proceeded to fill. They mobilized the engineering skills needed to build a small-scale unit capable of supplying 4,000 spindles with the high quality cotton ready for spinning.

The pre-spinning processes are, in sequence, first cleaning the raw cotton; then carding (combing) it to separate the fibres; drawing it to make the fibres parallel; and the final operation produces a thick thread with a slight twist, called rovings, which are taken over by the spinner and spun into threads of the required thicknesses.

By eliminating a number of sophisticated labour-saving devices the ATDA have been able to construct a pre-spinning unit of which the machinery cost, per spindle, is about half that required for the pre-spinning machinery of a large mill.

The first completed pilot plant, at Ghazipur in Uttar Pradesh, now comprises a service centre capable of supplying high quality rovings to 4,000 spindles dispersed among some 200 families living within a ten to twelve mile radius of the centre. The centre produced its first rovings in January 1979, less than a year after the site was selected. Experience with the pilot plant revealed the need for further modifications and improvements; these are now in hand in collaboration with the Shirley Institute in Britain, through the offices of the I.T.D.G.'s Industrial Services unit.

The centre's 'package' includes the purchase of raw cotton, credit arrangements enabling local spinners to buy the machines on hire-purchase, and technical and marketing services.

The centre also provides the technology needed to upgrade traditional weaving practices. One of the most time-consuming operations for the weaver is warping, the preparation of the beam or long roller which carries the tension threads on his loom. Hand warping is slow, and the result is often of uneven quality; hand sizing suffers from the same defects. By mechanizing warping and sizing, the weaver's productivity and the quality of his cloth are both enhanced. With this service, which the centre now provides, a weaver can expect to increase his earnings by up to 15 per cent.

Post-weaving treatment of cloth, for which the ATDA are now searching for improvements, involves dyeing, bleaching, callendering and sanforizing. Other technical improvements that are in the pipeline include a loom which could cost as little as 10 per cent of a conventional power loom. It is simple to operate and can be adapted to operate by foot treadle, a cycle attachment, or a motor.

The technology of decentralized, small-scale spinning, and of upgraded weaving, therefore, now exists. Spinning and weaving have again been reunited under one roof – that of the weaver's cottage – in such a way as to make possible competitive levels of quality and productivity, and higher income: taken together, the technologies that have already been proved by the centre can increase weavers' net incomes by up to 60 per cent.

Service centres along similar lines – run as commercial operations, as an integral part of the industry or community they serve – are obviously capable of wider application. The single-industry centre may be appropriate when there is a geographical concentration of a particular industry, as is generally the case with handloom weaving. But some might service a number of activities, such as carpentry, blacksmithy, building materials, water supply and sanitation, for a particular locality. How such centres can be brought under the ownership and control of the people they serve is a question yet to be answered, and one that deserves close attention and practical experiment.

The wool industry, and especially wool spinning, is another part of the textile industry in which the ATDA is working. The conventional, large-scale technology of wool spinning has been developed mainly in response to the need to process wool from sheep specially bred to produce wool of high quality. But in most developing countries, the wool is generally of poor quality. Selective breeding can raise the quality, but this will take a long time. So ATDA are now starting to do for the small-scale wool industry what has already been done for the large, namely adapt the technology to the raw material; in this case, the coarse wool produced in the hill areas of India.

In the hills of Uttar Pradesh, cottage spinners produce woollen yarn using simple hand and treadle operated spinning machines or charkas. The yarn is of variable quality, however, because of the practice of drafting fibre and inserting twist manually. The result is yarn which varies considerably throughout its length in diameter and twist content and, hence, strength. Further productivity of spinners on the traditional charkas is so low that the average income earned from a day's spinning is only Rs 1–2.

A direct result of the poor quality yarn is that the productivity of cottage weavers suffers. Cloth marketability is also adversely affected. These factors have contributed to the decline of this important cottage industry.

On a recommendation from the International Wool Secretariat, I.T.D.G.'s Industrial Services unit therefore commissioned Peter Teal, a British spinning expert, to develop a spinning device incorporating mechanical drafting and twist insertion. Design work was based on the 'two-for-one' twisting principle – producing two turns in the yarn for each revolution of the spindle – and the result was to raise rates of yarn production from 300 yards per hour to over 500 yards per hour per spindle. Single spindle prototypes have been built and tested in the U.K. and in India. The latest models produce yarn of mill-standard quality.

This collaborative project will field test up to fifty of the single spindle models using mill-produced slubbing. Simultaneously a multi-spindle version will be developed, incorporating two, three or four spindles. The preferred model of the multi-spindle version will also be extensively field tested. The device can be made from components readily available in India and the small amount of machining necessary can be accomplished on the small lathes commonly used in rural workshops.

Once adopted by spinners, it is estimated the device in two-spindle form will raise incomes from the present Rs 1–2 for a full day's spinning to Rs 6. The significant improvements to yarn quality will improve weavers' productivity and make finished Khadi cloth competitive with mill-produced cloth.

For two other examples of the ATDA approach, and of the potential that exists for upgrading derelict village technologies, we can turn to leather tanning and shoemaking, and village pottery.

Leather tanning offers a classic example of the disintegration of rural industries in India. The treatment of animal products probably comes third in employment potential in the country, after agriculture and textiles. In pre-industrial India, the village tanner got his hides on a barter basis, the hide being his payment for removing the carcase. He tanned the hide in his cottage, in a bag containing tanning liquid made from the bark of local trees. There was a good market for the leather, for shoes and harness, and especially for the large leather bags that were used to raise the irrigation water from wells.

The leather tanned in this way served its purpose, but it was of poor quality as only about 65 per cent of the hide was penetrated by the tanning liquor; and when modern technology came in, with an 85 per cent tanning efficiency, the village tanner lost his shoe leather and harness markets. Waterbags disappeared as mechanical water pumps took over. Meanwhile the demand for hides by the city-based tanners raised their price, and left the local tanner with access only to the poorest hides.

Do low-cost tanning materials and methods exist that could revive the rural tanning and leather industry? If not, can they be discovered? At village level, the current practice is to soak one or two hides in tanning material in a concrete lined pit. The first need is to raise the tanning efficiency – the extent to which the tanning liquid penetrates the hide – from its present level of 65–70 per cent to 85 per cent. This would enable the village tanner to supply the local market with good leather, and also to sell hides to the urban market. The next step must be to find ways of re-establishing shoe making as a rural craft.

Another rural industry that has steadily declined over the years is pottery – the traditional red clayware or terracotta that at one time furnished all of India's domestic utensils. It has been progressively replaced by glass, ceramics and metal, with the result that the number of village potters has declined by about two-thirds over the past seventy years, to just over one million today.

Whiteware pottery has already been cited as an example of a large-scale technology that has been successfully decentralized, at least to semi-urban areas and large village clusters. Unlike the mini-sugar industry, however, whiteware manufacture has not yet developed its own momentum and it remains to be seen how far centres such as at Chinhat can be established in selected rural locations. But as things stand, whiteware pottery is too expensive for the great majority of village potters to take up, and their needs have to be tackled in a different way. The most likely prospect is that of developing new products for red clayware, and especially products for which there is widespread demand. One idea is to promote the use of red clayware to make disposable cups and plates for serving tea in India's thousands of railway stations, and on trains; this could replace the unhygienic crockery now used with an attactive, cheap and bio-degradable product. (The use of red clayware cups is already widespread in roadside cafés in many parts of India; it eliminates washing up, and pilfering.) Experiments are now also being made with the use of red clay for decorative floor tiles, and frescoes, for housing, and with red clay pipes for sewerage and small-scale irrigation schemes; by putting low-cost pipes underground, instead of using surface channels, precious land can be saved for cultivation.

For red clay water pipes to be made cheaply and in large quantities a mechanical press is essential, and the ATDA are now searching for an efficient, low-cost pipe-extruding machine, preferably one that can be operated either manually or from a power source.

The more efficient firing that would be needed to make some of these new products, as noted earlier, already exists in the form of the family size kiln that was developed for small-scale whiteware production. Such kilns can be used for both whiteware and red clayware, and they will be tried out in a new service centre that ATDA are planning to establish at Sevapuri, a rural industries training centre near Lucknow.

These examples show, beyond any reasonable doubt, that rural industries *can* be revitalized, but only by the systematic application of the principle: find out what the people are trying to do, and help them to do it better. There is no natural law that decrees that people in rural areas cannot develop their skills, make products of the highest quality and earn a decent living.

Neither is it pre-ordained that only traditional industries should exist in the rural areas. No one has any idea of the potential that may exist for scaling down and decentralizing industries that are now highly concentrated into large, capital-intensive units. This is for the simple reason that with a few exceptions, no real effort has been made to scale down and decentralize them. One of these exceptions is in sugar manufacture, where, as we have already seen, mini-plants are (though only in India) an established part of the industry. Cotton spinning is well on the way. Two other industries that are promising candidates for scaling down are paper and cement making. ATDA's work on scaling down the technology of modern paper making is still in its exploratory stages; work on the mini-cement plant is well advanced, and looks very promising indeed.

Paper making, on ATDA's diagnosis, needs two scaled down technologies if it is to be decentralized on a competitive basis. One is a mechanical paper-lifting machine.

Traditional handmade paper in India is centred on about a dozen communities, such as Sanghaner near Jaipur and Kalpi in Uttar Pradesh. They make very high quality paper for traditional account books and legal documents. There is also a more modern section of the handmade paper industry which is more widely distributed, comprising some 200 units, each of a quarter of a ton per day capacity, set up by the Khadi Commission. They make a limited range of papers, such as writing paper and folders. They are also mostly working well below capacity, partly owing to shortages of raw material, but chiefly because of problems that are inherent in the process of hand-lifting itself. This is done by two men, who dip a rectangular sieve into the slurry (pulp and water) tank, and lift it out with a layer of pulp on the surface. The water drains off, and the wet sheet of paper is then dried in the sun. The process is slow, the quality of paper variable, and there is a lot of waste: about one-third of the output has to be recycled.

There is room for improving the equipment used in handmade paper; but to improve the quality and range of papers produced, there has to be mechanical not hand-lifting. The ATDA team are currently exploring the potential of a unique one ton a day paper-lifting machine manufactured for its own use by a large paper making company near Poona.

To make a small-scale paper industry competitive it will also be necessary to ensure its raw material supplies, especially of long staple fibres for making high quality pulp and paper. Small-scale – ten ton a day – plants have been successfully developed in India for mechanical pulping. But mechanical pulping produces only short-fibre, low quality paper. The question is, how far can chemical pulping be miniaturized? On present knowledge it seems that it might be possible to scale it down to thirty tons a day. If that is so, a central pulping unit – the service centre concept again – could serve a cluster of small-scale paper plants, including both one ton a day mechanized units, and small handmade plants.

While the scaling down of mechanized paper and pulp making has only just begun, the development of mini-cement plants is almost complete. Cement manufacture is one of the industries that has concentrated into larger and larger units during the past thirty years or so; plants making up to 3,000 tons of cement a day now represent a conventional technology. The exception is China, where the number of mini-cement plants – of about twenty-five tons a day – has risen from about 200 in 1965, to about 2,800 in 1975. These mini-plants employ about quarter of a million people in the rural areas, and account for more than half of China's output of cement. It seems, however, that they make cement of lower and more variable quality than do the large plants making standard Portland cement.

In India, experimental work on a mini-cement plant was started in the 1960s by M. K. Garg, when he was with the Planning Research and Action Institute. They built a twenty-five ton a day plant at Mohanlalganj, near Lucknow, but although it produced 10,000 tons of cement for which there was a ready market, only about 80 per cent of the output was up to Portland cement standards, and the project was shelved.

Since then, technical investigations made by ATDA, and the

Cement Research Association of India, have revealed the causes of the trouble, and a twenty-five ton a day plant making standard Portland cement is now a commercial possibility. The original plant at Mohanlalganj is being modified and refitted by ATDA with technical assistance from Germany (where there is long experience of the appropriate technology). A plant on similar lines is said to be undergoing its test runs in Tamil Nadu in southern India, and the Cement Research Association is going ahead with the development of fifty to 100 ton a day plants.

Mini-cement plants clearly illustrate the advantages of decentralizing large-scale industry by means of small-scale capital saving technologies. They make available an essential building material near to the point of use, and save on packing and long and expensive transport hauls – people can collect their own, in their own transport. They can use small deposits of limestone which are too small for use by a large plant (and there are many such small deposits all over India). They provide employment in rural areas – and the capital investment per ton of output is about half that for large plants. They can also be built and put into operation much more quickly.

Mini-plants can produce cement at relatively lower costs chiefly because they can use a more energy-efficient process. Basically, cement is made by grinding limestone and clay to a fine, consistent mix, heating this mixture to a temperature of about 1,400°C, and finally grinding the resultant clinker down to give the finished product. All the processes of grinding, mixing and so on can be efficiently miniaturized without much modification; but in the large-scale (rotary) plant, the heating is by radiation, while in the mini-plants (which use a vertical kiln) the heating is by convection and conduction. This uses significantly less energy per ton, and in addition the mini-plant can use cheaper forms of fuel, such as coke breeze and coal dust.

It is reasonably certain that within a few years from now, mini-cement plants will have become a conventional technology in the cement industry.

Work on mini-cement plants is a good example of the co-operation that can exist between action-oriented A.T. groups in the field and resource-oriented A.T. organizations in industrialized countries. ATDA and I.T.D.G. have worked together to carry out the techno-economic studies, to locate the German expertise and get the necessary financial backing from the German government and from Appropriate Technology International in the U.S.A.

But even when a new technology has proved its commercial success, ATDA argues, it must continue to get a back-up service of research and development if it is to remain competitive and responsive to changing market conditions. This kind of support service will not emerge automatically, because very few research establishments are geared to the needs of small industry or the rural economy.

In the case of industries where the technology needs scaling up, as in textiles, pottery and leather, this kind of support would be the function of their respective service centres. For mini-plants such as cement and sugar, once established, there is a similar need for continuing technical support, and ATDA's aim is to build up this capacity. Thus although the mini-sugar industry is now well established in India, there is still a lot of scope for improving its technical efficiency, but there is no research establishment working on its problems. All the development work done on mini-sugar plants was, in fact, done without any central government support. Yet nearly 40 per cent of the tax collected from large mills goes back to them in the form of research institutes, training facilities, pilot plant development, improvement of cane plantations, and so on. Now, with the support of the Uttar Pradesh State Council of Science and Technology, ATDA are undertaking new research and development work on mini-sugar plants. The development work is being done at a factory owned by the Gandhi Samarak Nidhi (which organizes Gandhian field work in India) and in collaboration with the Planning, Research and Action Department of Uttar Pradesh State, the research organization that was originally responsible for building up the mini-sugar industry's technology.

This means that after a lapse of several years, technical improvements will again be fed into small-scale sugar industry. The aim is to make good specific shortcomings in the mini-mill; the operations of cane crushing, to extract the juice; the evaporation of the liquid sugar before crystallization; and the process of crystallization itself.

Already significant improvements in the efficiency of cane crushing have been demonstrated, by substituting a screw-type expeller for the conventional rollers. A new type of evaporator could further reduce sugar losses. Compared with the large mill, the mini-plant loses a lot of sugar in the molasses that is left after crystallization, but now a process (ion exchange) is being tried out which could recover most of the sugar from the molasses – and this process is much easier to apply on a small scale than it is on a large.

If these three new technologies can be successfully introduced they could, by themselves, raise the total sugar recovery of the mini-mill (per ton of cane crushed) to at least that of the large mill, and certainly keep the small-scale sugar industry well abreast of any competition.

So far we have looked at ATDA's involvement in a variety of existing or potential rural industries. We now turn to the third wing of their programme, the improvement of domestic (home living) and community technologies; to raise the standard of living and the self-sufficiency of rural families and communities. Here they are concerned with subjects such as village power supplies from locally available and renewable resources, water supply and sanitation, food processing, social forestry, animal husbandry.

As with the industries programme, in each case analytical studies are made to reveal local needs and resources, what technologies already exist, what gaps can be filled by research or development, and how that can be done; and after that to set up, generally in collaboration with research or field organizations, pilot ventures to test the validity of the technology or package of innovations arrived at. Here are some examples.

Their survey of work on biogas, for instance, has led to the conclusion that a new research and development effort is needed. There are probably thousands of organizations and groups in the world working on biogas, that is, methane production from animal excreta and vegetable waste, and in India this work has been going on since the 1940s. Yet the basic problem still remains: about 80 per cent of the rural families in India own three animals or less – and no biogas system has yet been devised that enables these, the poorest, families to meet their cooking needs from the amount of biogas that can be extracted from the dung of three animals. By burning the raw cow dung they can get just enough heat to do their cooking. Consequently most of the cow dung produced is not returned to the land but is burnt, with very serious consequences for the fertility of the soil.

There is also a problem of cost. The cost of a family size plant (about Rs 3,000) producing 100 cubic feet of gas a day is quite beyond the reach of most villages.

In the standard biogas plant, cow dung is mixed to a slurry and fed into a cylindrical tank, usually sunk into the ground, with a floating domed steel lid that acts as a constant pressure storage unit. The gas is generated by micro-organisms that inhabit animals' intestines, and they operate only within a fairly narrow temperature range; they work best at a constant 35°C. The spent slurry makes a rather better fertilizer than the original raw dung.

Recently many advances have been made in the design of biogas plants, especially in China. The Planning Research and Action Department of Uttar Pradesh State, which is building 10,000 biogas plants in U.P. alone over the next few years, are experimenting with Chinese models of 100 cubic feet and upwards, in which metal is totally replaced by brickwork and mortar. This has cut the cost of a family size unit by nearly half, to about Rs 1,500–2,000; but that is still far beyond the reach of most families and of course the basic problem remains: most people have fewer cows than needed for self-sufficiency in cooking gas.

There is undoubtedly scope for more community plants. The first pilot community plant installed by the PRAD was at Lal Kapurwa village, U.P. This is a community of some thirty families who formerly used kerosine for lighting and cow dung and crop residue for cooking. The community plant comprises two tanks of 25 and 40 cubic metres (28 and 52 cubic yards) capacity, and from these gas is piped to each houshold for cooking and lighting, and street lighting is also provided. The gas is also used to drive an engine (which runs on an 80 per cent gas, 20 per cent diesel mixture) which pumps water to a 10,000 gallon overhead tank to supply the village. When not pumping water, the engine can drive a flour mill, a thresher and a chaff cutter.

One of the aims of this pilot plant is to discover whether the costs can be met by the community on a continuing basis. Economics is almost certainly not the limiting factor in the case of community plants, the future of which really depends upon how many communities are both able and willing to run them.

Whether or not community plants can work, ATDA argues, what is needed is a substantial increase in the efficiency of the process. Currently, biogas plants yield about 2·5 cubic feet of gas per 2 lb. of dry cow dung. By heating or insulating the tank, and agitating the slurry, this figure can be improved upon, but invariably at greater cost. Ways must therefore be found of first, significantly raising the gas yield per lb. of dry matter, and secondly speeding up the cycle, the rate at which gas is produced.[4] Only then is there any real prospect of reducing the size and cost of the plant enough to bring it within the reach of the majority of rural families.

To achieve either or both of these objectives – more gas per pound of cow dung, more quickly – requires new work on the microbiology of the process, about which relatively little still appears to be known. Here is a challenge to universities and research establishments everywhere. Meanwhile ATDA have promoted work along those lines at the Central Drug Research Institute in Lucknow.

Another unconventional energy source that could make a useful contribution to village development is mini-hydro. In the hill areas, many small streams with falls of five to 100 feet have already been harnessed to small-scale hydro systems thanks to the development work of the Jyoti company, a large engineering firm in India. These produce between fifteen and fifty kilowatts of power. But there are many smaller perennial streamlets with falls of about twenty feet. A few are already used by local people by means of a traditional wooden water wheel which provides direct mechanical power for small sawmills, or grain grinding. For these there is as yet no off-the-shelf technology, and if one could be developed, the basic power needs of most of the hill communities could be met very quickly and at very low cost per kilowatt. The amounts of power that would be generated are small, between five and fifteen kilowatts per installation. But then five kilowatts is enough to pump water sufficient for 200 people in three hours or so, as well as to provide lighting at night.

The Association's work is done in collaboration with many other organizations: with government and university departments of research; with the Uttar Pradesh State Council of Science and Technology in connection with work on sugar; with the Appropriate Technology Cell, Ministry of Industry, which has commissioned them to do a survey of very small industries with starting up costs of less than Rs 10,000; with equipment manufacturers of sugar, rice and oil expelling plant, and with the banks to arrange for loans to village spinners. There is very close but informal liaison with the Uttar Pradesh State Planning Commission, and they get a great deal of support from the Gandhi Samarak Nidhi.

For the purposes of identifying specific needs in rural communities, and testing out new techniques and methods, ATDA rely upon experienced field organizations. One is Sevapuri, near Lucknow, which specializes in the training of rural artisans and workers in crafts such as leather, pottery, weaving, biogas, oil extraction and fruit processing.

Another centre of outstanding quality is Gandhigram near Madurai in south India. They have been running for three decades, with a steadily growing emphasis in recent years on ways of increasing the incomes of the rural poor.[5]

They now employ over 100 people in their rural industries programme. Five small service centres have been set up, each serving between eight and sixteen villages. A typical service centre has a health clinic and also small dairy processing, soap making and spinning and weaving units. These supply improved equipment and raw materials to the villagers and train operatives in new skills. A service centre processes the milk into butter; and simply by making and selling butter, they can pay the villagers 6 per cent more for their milk than the villagers could get by selling milk to urban milk depots. The centre returns the buttermilk residue to the villagers free – thus improving both the villagers' incomes and their nutrition.

The activity that is being generated at village level is reflected in the questions that reach ATDA through Gandhigram headquarters: How can you improve the quality of village tanned leather without putting the process beyond the reach of the villagers? What would be an appropriate goat for crossbreeding with local goats to raise their milk yield? How to increase the shelf life of a malted drink made from local grains without impairing its nutritional value? What simple bamboo splitting machine or tools exist that can produce a wide variety of bamboo strips and fibres? Is there a simple and effective device for callendering handwoven cloth? (At present this is done by hand-beating the cloth.) Can a simple cooling chamber for milk be devised, suitable for humid climates, that will reduce temperature by at least 5°C? Are there efficient, small-scale cheese making units to cope with about 55 gallons of milk a day? Can milk powder be made on a similar scale; and how can butter, cheese and milk powder be packaged at lowest cost?

These are the kind of questions to which India's appropriate technology centres must find practical answers. There is no doubt that Gandhigram, and other centres working at village level, can put these answers to very good use, as tools in the hands of the poor.

Pakistan

Pakistan's Appropriate Technology Development Organization was formed in mid-1974, on the recommendation of an I.T.D.G. team, headed by Schumacher, which had visited Pakistan some six months earlier. Although from the outset it was intended to be an autonomous group with government funding, in its early years it was kept going with the support of the commercial banks and the Lahore Chamber of Commerce, and they continued to help the new organization in various ways up to 1976. Gradually more official funds became available. In 1978 ATDO was constituted as an autonomous body, like the Pakistan Council for Scientific and Industrial Research, under the administrative control of the Ministry of Science and Technology.

By the end of its fourth year ATDO could report that work was proceeding on more than thirty projects (see Table 1).

TABLE 1: Examples of Existing Projects

Food and Agriculture	*Waste Utilization*
Dehydration of fruits and vegetables	Cement from rice husks
Screw type can crushing	Paper from banana stems
Undersoil irrigation	Gobar gas plants
Coconut plantation	Insecticides from tobacco waste
Village level food processing (sugar cane and oilseeds)	
Energy	*Agro Implements*
Mini-hydro systems	Earth moving implements
Windmills	Oil expeller
Water turbines	Multi-hopper seeder
Solar energy	Paddy drier
Small-Scale Industry	*Health and Habitation*
Utilization of local iron ores	Low-cost housing
Manufacture of chalks, candles at cottage level	Low-cost primary schools
Handmade matches	Village water and sanitation
	Mobilization of technicians' skills

Several of these technologies have reached the early stages of field extension, notably mini-hydro, low-cost housing, and gobar gas. Work on these, and on other technologies such as improved village level sugar cane crushing and processing, under-soil irrigation and agricultural equipment, was being done in collaboration with the Universities of Peshawar and Lahore, with government research establishments, and with private firms and individuals.

The most convincing achievements have been the development and introduction of mini-hydro systems and partially prefabricated low-cost housing. After some three years of development and testing of mini-hydro by private individuals, and at the University of Peshawar, full-scale field extension was started in 1977. Three villages in the Swat district have been electrified, and a fourth was well on its way by early 1978. A similar project is going on in Sultanabad, in Gilgit.

The mini-hydro units comprise a ten-kilowatt electric generator, and a water wheel made of steel or plastic. The civil work required – digging the power channel and building the power-house – is done by the local people. In Sultanabad they have organized themselves into a co-operative, and other villages are following suit.

The savings in capital costs per kilowatt installed are impressive. If the government provides the generator and the water wheel, the cost per installed kilowatt borne by the state is Rs 1,000 – compared with Rs 35,000 per kilowatt if the work is done by conventional methods; and in the case of the mini-plant, maintenance is also done locally.

In the areas where there is potential for mini-hydro, the commercial banks are arranging to finance the purchase of the necessary equipment by village co-operative societies. The ancillary work, such as distribution and house-wiring, is done by the people themselves through their associations.

As the generators have to be imported, ATDO commissioned work at the University of Technology, Lahore, to develop local capacity for their manufacture. The aim is to produce locally five, ten and fifty-kilowatt generators.

ATDO has also sponsored successful work on low-cost housing, using hollow blocks and prefabricated concrete battons and slabs for roofing (to minimize the use of steel and of very scarce wood). This work was done in collaboration with the Building Research

Station, and a local Karachi contractor who donated part of the development costs. Co-operative housing societies in Karachi are building 7,000 houses based on these designs. Subsequently ATDO designed a smaller house using similar modes of construction, which could cost as little as Rs 6,000 (£300). As people in Lahore were sceptical about this, it was agreed with a leading journalist that ATDO should build a house on his plot; if he disapproves of it, it will be removed, but if he accepts it, he will meet the cost.

A major project in the food and agricultural sector is the Village Level Food Processing Programme. This was started at the end of 1977, and it seeks to identify, develop, produce and disseminate village-scale technologies for processing sugar cane, oil seeds and rice bran. The Denver Research Institute, Denver, Colorado, are the prime contractors for this project. The technologies identified by the Institute are currently being tested and improved by government research laboratories.

In support of low-cost housing, the ATDO has also been taking an active interest in work on cement from rice husk ash. This technology is being developed by the Pakistan Council for Scientific and Industrial Research on lines similar to these being followed in India. The ash, mixed with slaked lime, gives a product with many of the qualities of cement. A mill is being set up in a rural area, to test the economic and social acceptability of the material.

Like some of its Indian counterparts, ATDO has also been working on ways of raising the efficiency and lowering the cost of gobar gas production. With the help of technical data from India and China, they have been promoting the use of plants that employ a concrete dome in place of the steel gas holder of the conventional gobar gas unit. The dome is less costly, and can be made locally. After a number of the new units had been successfully operated, the commercial banks agreed not only to provide finance for such plants, but also to disseminate the necessary technical knowledge from the banks' branches. Meanwhile ATDO started fabricating portable shuttering for the dome type cover, and other technical devices to aid construction in rural areas. The aim is to make these, too, available through the banks' branches, along with instruction books and drawings.

At a more sophisticated level, ATDO recently started preliminary research and development work on a mini-iron and steel plant, the object being to develop a small plant that could be locally fabricated, and which is capable of using low quality iron ores. Development work is being done with the collaboration of the Pakistan Minerals Development Corporation, and support from the U.N.D.P. During 1978 experiments were going on in a test plant at Rawalpindi, but the results are still far from positive.

One of the features of the ATDO programme is its emphasis on securing the widest possible public participation in the development of appropriate self-help technologies. Thus the whole concept of the mini-hydro programme is based on people's participation, on the sensible principle that what people have made for themselves, they will maintain. The pioneers of the improved gobar gas plants were not high-flown scientists, but two local farmer-craftsmen. The successful adaptation of a bullock-drawn earth scraper (from designs produced by VITA) was the work of a technical college blacksmith. Appropriate technology can flourish, ATDO argues, only if large numbers of local craftsmen are involved, and are given opportunities to develop and apply their skills.

Sri Lanka

The Sarvodaya Shramadana Movement (S.S.M.) seeks integrated rural development based on traditional and cultural values. People's participation is the cornerstone of the movement, which was founded in 1958. Buddhist philosophy has a strong influence on the S.S.M., and they look upon their work in the villages as the creation of a suitable livelihood rather than the creation of jobs through economic action. The Movement's founder and leader is one of the great men of this generation, A. T. Ariyaratne.

The first task of the S.S.M. is to make villagers aware of the reasons for their present predicament and convince them that economic regeneration must be preceded by a restoration of social values and relationships. Over 800 villages are now at this stage.

The second stage involves the organization of groups of different ages and occupations and with varying responsibilities. Some 1,900 villages are at this stage.

In the third stage, a village development plan is drawn up by the village council with the help of S.S.M. workers, and about 250 villages are now working to their plans. To support all these programmes the S.S.M. has set up some 1,100 pre-schools, over 120 extension centres, and fourteen development educational centres.

Over one million people are involved daily in Sarvodaya activities and in addition there are some 5,000 village level workers.

Experience of government rural industrialization programmes has shown that it is not enough merely to establish village industries. The S.S.M. believes that education must be introduced in management and production, in marketing, and in technical services.

Education is based on the traditional apprentice system as practised in the Kandyan period (1487–1857). Apprentices are also trained in the theoretical aspects of their arts or crafts, and have lessons in simple record and book-keeping. New skills, too, have been introduced.

Marketing is through community shops, where goods made in the villages are for sale, and they are also supplied countrywide through a S.S.M. marketing organization.

As far as possible, village industries use locally available raw materials and renewable energy resources. The criteria for buying any foreign equipment are that repairs can be carried out within Sri Lanka, and that spare parts can be made locally. The movement concentrates on three types of industries: carpentry, blacksmithy (agricultural implements, tools, wheelbarrows, etc.), and food processing (vegetable oil and cereals).

The average size of the industries promoted is ten workers per unit, and the average cost per unit is Rs 4,500 (about U.S. $300 at present levels). Technical and managerial support services help not only with the repair and maintenance of equipment, but also with the keeping and use of records, and quality control.

Village industries produce for a local market which activates the village economy and increases the standard of living.

Africa

Ghana

The Technology Consultancy Centre at the University of Science and Technology, Kumasi, was started in January 1972. For some time before, a group of faculty members had been giving technical advice to small local manufacturers, and through this work had established contact with I.T.D.G. During 1969–70 the Group, with the collaboration of the Inter-Universities Council of Britain and the University of Edinburgh, helped to get the Technology Consultancy Centre launched as an official part of the University of Kumasi's activities.[6]

The T.C.C.'s purpose is to make available to the public the technical and scientific expertise of the University, and to promote the industrial development of Ghana; most of its work is on small-scale industries.

The T.C.C. is run by a staff of only seven professionals in the fields of engineering, agriculture and industrial art; but it draws on thirty to forty members of the University faculty for planning and executing its projects and for consultancies; and employs more than fifty technical staff in production units on the campus. Its work falls into three broad categories: technical and commercial advice to industry and government; development and testing of new products; and the commercial operation of production units on the campus, sales from which account for nearly half of the Centre's total income.[7]

Many small manufacturers come to the T.C.C. to get their products tested or analysed before they go into production, or to ensure that they will meet the requirements of the Ghana National Standards Board. Others are directed to consult the Centre by banks to which they have applied for loans to start or expand a business, others again want advice on the supply of raw materials, or on choice of equipment. Thus over the past few years the T.C.C. has advised on the manufacture of an extraordinary range of products: gunpowder, rubber mouldings, wood and coconut charcoal, leather goods, envelopes, sugar, chalk, kaolin, tonic drinks, jams, and brass castings; and it has done chemical analyses of soap, glue, bleach, alcohol, latex fluid, cassava starch, seashells and caustic soda. Many requests are passed on to appropriate members of the University faculty, who take on the work as consultants, charging fees to those who can afford to pay. In recent years consultants from the Faculty of Engineering have worked on the installation and commissioning of a carbon dioxide welding machine, the restoration of the air conditioning and refrigeration plants of the biggest hospital in the country, the design and manufacture of cheap feeder tubes for rabbit farming, the design of metal stamping dyes, the repair of a laundry steam press and the repair of a large wood-fired steam boiler at a plywood factory.

But other requests have led to the development of new products and industries. Among the industries for which the Centre has developed and helped to introduce appropriate technologies are glue, soap, craftwork, and animal feedstuffs. The first three are excellent examples of how the intelligent use of technology can upgrade and make more profitable a variety of traditional activities. In the case of animal feedstuffs the technology was an innovation, with the added bonus that it brought into productive use raw material that had formerly been discarded as waste.

In 1972 a small manufacturer of cassava starch for local laundries asked the T.C.C. to help him to make paper glue of better quality than the traditional glue, which was made by mixing cassava starch with water; glue made that way has a short life, and is not re-wettable.

The University's Department of Chemistry provided the answer, by adding a non-toxic fungicide as a preservative, and chemicals to render the glue re-wettable. The result was a glue every bit as good as imported products.

The T.C.C. also advised the manufacturer on how to apply for a bank loan, and to run some market tests; and the equipment needed to start production was made to the manufacturer's order in the T.C.C. workshops. That was how 'Spider' glue was launched on the market. More glue makers then started up, but the original manufacturer is still doing good business, employing eight people and producing 8,000 bottles of glue a month. By 1975 this glue was meeting all of Ghana's needs, and although there is still an imported content in the form of plastic for making the glue bottles and some additives, the new industry is saving the country over $100,000 (at mid-1978, $2.30 = £1 sterling) a year, making better use of local raw materials and creating jobs and income.

In that instance the T.C.C.'s participation was relatively modest, consisting chiefly of technical advice. In the next case, soap making, the problems were greater.

As with glue, the T.C.C.'s work on soap started in response to requests from traditional soap makers to improve the quality of their product. In Ghana there are many soap makers working on a very small scale, each making about eighty one-kilo (two-pound) bars of soap a day (about a ton a week), using fourty-four gallon oil drums for boiling the alkali-oil mixture. This soap is of low and variable quality. At the other extreme, there is a very large-scale soap factory in Ghana. Owing to raw material shortages it works at about 20 per cent capacity; but even if it worked at full capacity it could not meet more than half of the country's demand for soap, and as it is it relies on mostly imported raw materials. So on grounds of import saving, and supplying more and better soap to the market, there was a need to upgrade the technology, to find something 'intermediate' between the huge factory and the forty-four gallon oil drum. That is what the T.C.C. set out to do.

Experimental work was started in the Centre's own workshop, with technologists from different faculties working with a local soap maker. This was followed by full-scale trials in a prototype plant built on the university campus. Within a year a satisfactory formula for good quality bar soap was ready, and the T.C.C. unit, with a capacity of about 500 bars (about half a ton) of soap a day, was in business, selling all it could produce.

When caustic soda – one of the main ingredients of soap – became hard to get, a member of the Chemistry Department designed a small (200 lb. per day) caustic soda plant; this uses slaked lime from a neighbouring factory and imported soda ash, and produces caustic soda at about one-third of the prevailing market price. (The plant proved to have other uses too, such as making insecticides and bleaching fluid, so some more were built.)

While continuing in production, the prototype plant was used to train workers for a full-scale pilot plant which the T.C.C. was building off-campus as a demonstration unit, and also for five entrepreneurs who had placed orders with the Centre for soap and caustic soda plants. The prototype plant was later renovated and sold.

Problems of quality control continued to arise, so in 1975 a soap expert was brought in from India. His recommendations were immediately effective. He showed that when electrical heating was replaced by wood-firing this cut the initial cost of the soap boiling tanks by 70 per cent, halved fuel costs, and reduced the process time by 10 per cent. He also introduced better methods of quality control, and identified locally available white kaolin as the best additive to the soap.

Severe shortages of raw materials and especially of palm oil have set back this programme, but at least four entrepreneurs and the T.C.C. pilot plant continue to operate profitably. A large-scale programme of palm oil planting is going on in Ghana, but the trees take five to six years to mature; so the T.C.C. is working on more immediate solutions to the problem. It has already developed a hand-operated screw press which improves the efficiency of oil extraction from the available supplies of palm fruits, and is developing a low-cost unit for distilling perfume (another ingredient of soap which has had to be imported at high cost). The most promising line of approach to ensure adequate oil supplies is of course to discover substitutes for palm oil. Tests using castor oil and physic nut oil show encouraging results, and those plants take only five to six months to fruit. A ten-acre plot has been planted with these nuts to test out their commercial possibilities.

The same kind of systematic attention to local needs and resources has enabled the Centre to improve the economic environments for local craftsmen in the Kumasi region, such as weavers, glass bead makers and brass workers. These crafts are generally centred on a group of villages, and most of the craftsmen are also subsistence farmers and woodcutters.

In the case of the glass bead makers, of whom there are some 2,000 in the Ashanti region, the T.C.C. has been able to help the craftsmen to upgrade the quality of their product and also to expand their market.

Strings of glass beads are worn by all Ghanaian women (and some men). There are strings for wrists and waists; there are some sixty different designs, and as each has a meaning, different strings of beads are appropriate for different meanings and occasions. There is, in short, a ready market for beads.

The beads are made from recycled bottle glass, and fired in clay moulds in a wood-fired kiln. Having been asked by a group of bead makers to help them with marketing, the Centre workers very quickly gained the confidence of the craftsmen by finding a way of greatly improving the quality of the beads, and lowering costs into the bargain. They found that the beads were coloured by grinding down very expensive beads imported from neighbouring countries and mixing this material with the recycled glass. Instead, the Centre imported ceramic glass colouring, which gave greatly improved results and also cut the craftsmen's production costs by 25 per cent. At the same time the bead makers were helped to form co-operatives, through which they can get space in public markets, secure import licences for colouring material, and become eligible for credit to expand their production. Meanwhile, the T.C.C. are identifying and experimenting with locally available colouring materials.

Having found a source of real technical help the craftsmen make good use of it: T.C.C. has had over 300 enquiries from bead makers during the past few years, and is now experimenting with polishing machines, and improved moulds and kilns, to further improve the quality of the beads, and raise the productivity and the incomes of the rural craftsmen.

In the case of brassworkers the Centre is trying to widen the range of products that could be made by the craftsmen. There is a long tradition in Ashanti of casting brass figures by the 'lost wax' process. Many of the brass figures are complex, and all have a symbolic meaning exemplifying Ashanti proverbs.

The chief brassmaker of Kurofofurem village tried his hand at making brass parts for a water pump, and bushes for a rice thresher, and found little difficulty in adapting his technique to the new designs. Other fittings could be made in this way by the village craftsmen. The Centre is now experimenting with different types of moulds and kilns, and using aluminium and light non-ferrous alloys as well as brass. In time these might well lay the basis for a sizable local plant for making small engineering and domestic components, all of which are currently imported into Ghana. (The lost wax method of making castings is commonly used by large firms in industrialized countries for making especially complicated shapes, such as for sewing machines.

A non-traditional activity successfully introduced by the Centre is the manufacture of animal food from brewers' spent grain. A local Kumasi brewery produces over 100 tons a week of spent barley, which they had been paying a contractor to remove and dump as waste. Now a local entrepreneur is taking fourteen tons a week of this protein-rich material and drying it for sale to local farmers. The grain is 75 per cent water when it leaves the brewery, and in this state it ferments within a day. If the moisture can be quickly reduced to 12 per cent or less it can be stored for long periods. The Centre's role was to design a hand-operated screw press which extracts most of the water quickly, and sun-drying completes the process. This animal feed is in very brisk demand, and the production costs are so low that the owner of the new business, now employing four presses and eight workers, makes a good profit by selling the grain at prices that are very competitive with other animal feeds.

Everyone thought his venture would collapse when another brewery installed a highly mechanized plant for drying spent grain (after waiting nearly five years for a licence to import the equipment from abroad). In the event, the large-scale rival was nowhere near competitive; owing to high capital and energy costs, its selling price is nearly 60 per cent above that of the small business. (The capital cost per workplace is more than thirty times greater in the large unit than in the small, even when the large unit has double manning on a three-shifts basis!)

In the cases we have looked at so far, the Centre's work has arisen out of requests by people seeking help, either to improve their existing technology or start something new to meet an obvious and existing demand. In several cases where the technology is unfamiliar or the market not yet developed, the T.C.C. has set up pilot production units on the university campus, and runs them as commercial enterprises. The aim is to prove that the technology in question is viable, and to provide training facilities, and thus to encourage local entrepreneurs to start up similar small units off-campus, with the Centre providing technical and other back-ups as required.

Production units of this kind have been set up for small-scale broadloom weaving, and the manufacture of nuts and bolts, both new technologies in Ghana. The textile weaving industry in Ghana comprises modern, sophisticated mills in the towns, and traditional weaving – the seven-inch wide Kento cloth – in the villages. When some local weavers showed an interest in using broadlooms to diversify their output, the Centre taught a local firm how to make broadlooms, ordered twenty looms from them and set up a weaving production unit on the campus. This was to demonstrate the commercial viability of broadlooms and to train weavers.

The production of nuts and bolts was started in order to alleviate a shortage of coach bolts among local builders of wooden lorry bodies; then it was discovered that almost all the high quality nuts and bolts used in Ghana were imported, and were very expensive. The production unit is seen as the start of an indigenous nut and bolt industry, in a country where the mechanical engineering industry is virtually non-existent.

Both these units have shown (though in the case of broadlooms not yet conclusively) that there is a profitable market from their products in the country, but as yet neither has been taken up on a significant scale. Some fifty broadlooms have been set up by people outside the campus, and two small manufacturers of nuts and bolts are in business. As both the pilot production plants have been going for six years, the rate at which these enterprises have spread must seem disappointingly slow. At least part of the reason must lie in Ghana's extreme shortage of foreign exchange and her dependence upon imports in both cases: of raw cotton, and of all machine tools. But at least the groundwork for these new industries has been laid and the units making nuts and bolts are the only ones in Ghana making high quality products from locally available supplies of steel.

While these two industries may be a bit ahead of their time for Ghana, the same certainly cannot be said of the Centre's plant construction unit. This has made almost all the equipment for the prototype plants set up by the T.C.C., and it has also made to order most of the plant for the soap and other industries being developed by local producers. It is now also making agricultural equipment, especially for use on small farms. Twenty pedal-operated rice threshers, based on designs supplied by the International Rice Research Institute in the Philippines, are currently being field tested, and when proved will be a useful tool for the small farmers who produce some 60 per cent of Ghana's rice. Another project that holds great promise for a timber-rich country such as Ghana is the construction of four pyrolitic converters. These plants are for converting sawdust, currently a waste product produced in very large quantities (estimated at about 25,000 tons a year), into three useful fuels; charcoal briquettes, gas, and combustible oil. Initially the oil will be used for firing a local brickworks. This project is run in collaboration with the Georgia Institute of Technology, U.S.A., which designed and developed the plant.

The Centre is now planning to set up several intermediate technology transfer units. These are intended to take its services right into the small workshop areas of surrounding towns. Each unit will comprise a workshop demonstrating products and processes developed or adapted by the T.C.C. to suit local conditions, but always using tools and methods and raw materials that are within the reach of local workshop owners and artisans. Each workshop will include blacksmithy, carpentry, welding and sheet metal working. They will, in fact, be small service centres for their locality, offering tools and machines on hire, initial stocks of raw materials, training and technical and commercial advice; in short they will provide the local person in a very small way of business with at least some of the facilities now enjoyed exclusively by large firms.

By any standards the achievements of the T.C.C. are remarkable, and all credit must be given to the University of Science and Technology for having launched the Centre ànd supported its work programme. Here are the main conclusions of an independent consultant who recently reported on the T.C.C. and its work:

All of the technologies established by the Centre satisfy most of the criteria considered desirable in an intermediate technology. The only significant exception is that some are dependent upon imported machines or raw materials. However, in Ghana's current situation this is unavoidable because some things simply cannot be made in the country itself. The T.C.C.'s projects have only used imported goods when absolutely vital, and it is hoped that this will be recognised by the Government so that its allocation of foreign exchange will concentrate on such goods.[8]

The technologies have the following characteristics:

Processes are simple. Some, such as weaving and the manufacture of nuts and bolts, require skilled or semi-skilled workmanship; these can easily be mastered and the Centre has already set up training facilities in its own workshops.

Materials are mostly familiar ones and locally available, though some have to be imported. The projects have the potential between them to achieve considerable import savings.

The enterprises can all be small-scale, employing between one and twenty people each.

The capital cost is low, and it is within the capacity of interested entrepreneurs to raise funds for the required investment.

Much of the plant required can be manufactured locally. Many of the items have been made in the Centre's workshop, but most could also have been made in the small-scale workshops in the informal industrial sector, and transfer of their manufacture could easily take place when development work on the designs is complete, as it already is in many cases. The manufacture of some equipment, such as the broadlooms, has already been transferred.

With the possible exception of broadloom weaving, the profitability of the enterprises has been demonstrated.

The technologies are all capital-saving. In comparison with the techniques that would be used in developed countries, where capital is cheap and labour expensive, the intermediate technologies require less capital and more labour, as is appropriate in a less developed country where capital is relatively scarce and labour cheap. As a consequence the potential for employment generation per unit of capital expenditure is very much greater than would be the case with Western technologies.

The Director of the T.C.C., Dr John Powell, has compiled a table which shows the capital invested and value added per employee for the T.C.C. projects (see Table 2 opposite). It also shows that for these seven projects, the capital required to create one new workplace ranges from about £20 to £850. But even the highest of these figures is still only about one-third of the cost per workplace of a typical Western imported technology.

By way of a summing up, here is a view from John Powell:

The work has often encountered difficulties and frustrations arising out of the declining economic situation, the reducing availability of raw materials and the lack of motivation on the part of many of the entrepreneurs when the Centre has attempted to help. However, the experience has also had a happier side which may be illustrated by the following stories which suggest that the long awaited 'spread effect' may be beginning.

AN INSTANT SOAP INDUSTRY IN THE SHADE OF THE OIL PALM

The T.C.C. has been making small-scale soap plants and train-

Group	Industry	Nos. Employed[1]	Capital investment in plant and working capital per employee (Cedis)[2]	Materials used per employee per annum (Cedis)	Average annual wage (Cedis)	Output per employee per annum (Cedis)	Value added per employee per annum (Cedis)
1	Traditional soap making[a]	3[d]	110	848	n/a	1,440	592
2	Palm oil extraction[a]	7	310	1,299	731	3,429	2,130
3	Broadloom weaving[b]	16	450	823	576[f]	1,471	648
4	Brewers' spent grain[c]	9	770[e]	12	750	1,760	1,625
5	Soap pilot plant	13	n/a	4,025	1,672	8,556	4,531
6	Steel bolt production	11	2,830	1,994	1,209	3,657	1,663
7	Plant construction	6	4,390	4,582	1,625	8,951	4,369

a Data based on 32 weeks' production a year
b Data based on 3 best months of 1977/8
c 1966 prices
d Self-employed
e Including buildings
f Low because trainees paid low wages
1 Numbers rounded up to include part-time workers
2 To nearest ₵10 (5.30 Cedis = £1)

General notes:
The traditional soap industry is included for comparison. It is not a T.C.C. project.
Group 1 is a traditional technology.
Groups 2, 3, 4 are man-powered intermediate technologies.
Groups 5, 6, 7 are mechanically powered intermediate technologies.

TABLE 2: Capital Investment, Value Added and Employment in T.C.C. Projects, Ghana, 1977–8

ing soap makers since 1973. Several plants have been established in different parts of Ghana, some of which have prospered but some of which have run into problems of obtaining regular supplies of palm oil. In all cases, the process of learning has been a slow one and many hopeful would-be soap manufacturers who expected to make fortunes overnight have become disillusioned when presented with the hard facts of the situation. In particular, many who said that they wished to enter the industry had no previous training or experience.

When Mr Korye entered the T.C.C. office and stated his intention to become a soap manufacturer, the staff wondered if here was just one more hopeful but ill-prepared candidate. 'Do you know anything about soap making?' he was asked. 'Oh yes,' he replied, 'I have been working for Madam Dansowa who uses a T.C.C. plant for making soap and caustic soda at her place in Accra.' So here was a soap maker who would not need three months' training at the T.C.C.'s Soap Pilot Plant at Kwamo, Ashanti. However, several had failed because they could not obtain palm oil and so the next question related to this vital issue. 'My mother and sister have an oil palm plantation and produce palm oil. They have guaranteed me a regular supply,' was the reply.

Now it was the turn of the T.C.C. to surprise its client. 'Can you collect your plant tomorrow?' (it was a Thursday). 'No, but I can call for it with a truck on Monday,' Mr Korye replied, and on Monday he came and paid for his soap plant and took it away to Abease in the Eastern Region to set up his instant soap industry. 'Abease' has the appropriate meaning of 'under the oil palm'!

HOME MADE SAW BENCHES

Several years had passed since S.S.I.S. Enterprises, a small carpentry firm, had imported their universal wood working machine with the help of a loan from Scottish War on Want. The T.C.C. waited patiently for other carpenters amongst the informal industries situated on 'Carpenters' Row' at Anloga in Kumasi to ask for similar help in obtaining powered tools. However, apart from some carpenters bringing their timber to S.S.I.S.E. for sawing or planing, no spread effect was observed. Then in June 1978, Mr Nuhoho called at the T.C.C. and asked for some help in making a saw bench equipped with an electric motor driven circular saw. He collected the necessary materials and constructed the prototype at the T.C.C. workshop assisted by the technicians of the Plant Construction Unit.

Some months later during a visit to S.S.I.S.E., it was observed that a saw bench was being operated by a neighbouring carpenter. On investigation it was found that the saw bench was one of four recently made by local welders and fitters following the pattern of the one which had been made at the T.C.C. workshop. Yet another saw bench was under construction in the S.S.I.S.E. Workshop. Over the years, it seems that the desire for a powered saw had matured and when the local product became available it found a ready market. By the end of 1978 some forty saw benches had been produced and sold to carpenters.

These stories show that a programme pursued steadily over a period of several years is likely to begin to generate its own spread effect which becomes almost independent of the original change agent. Implicit in this experience is a warning that projects in appropriate technology should not be expected to yield quick results and should not be abandoned prematurely. As with all processes of natural growth, it is very difficult to see any movement for a long period but when movement is observed, its pace is seen to increase rapidly. The T.C.C. has several projects which have been operating for more than four years. It is hoped that the experiences related above will encourage younger programmes to press on with their appropriate technology projects in spite of present frustrations and difficulties.

Kenya

Two organizations in Kenya that are promoting the development and introduction of rural technologies are the Village Technology Unit in Nairobi, and the Craft Training Centre (Village Polytechnics) Programme.

The Village Technology Unit was set up on a half-acre plot at Karen, Nairobi, in 1976, by the Youth Development Division of the Kenya Ministry of Housing and Social Services with the assistance of UNICEF. It is a centre for the demonstration, testing and extension of village technologies. Its special emphasis is on technologies that raise the quality of family and community life: home improvement and ways of reducing the work load on mothers; food production, conservation and preparation; and better water supplies.

The unit includes a workshop of the type used in Craft Training Centres in Kenya, equipped with manual wood-working and metal-working tools, including a wooden foot-operated vice and an oil-drum forge. Part of the workshop is a 'laboratory' area with weighing and moisture testing and temperature recording equipment; this is for testing and evaluating different ways of drying, heating and storing food. Hot water is supplied to the workshops from a solar water heater, to demonstrate the utility of simple methods of solar water heating for use in clinics, health centres and schools. The workshop also demonstrates construction techniques including earth and cement block making and the use of sisal pith for insulation.

The home improvement unit, inside a typical mud and wattle hut, includes a simple raised smokeless cooking stove, an oven made of oil cans, a hot-box cooker and an improved water cooler and filter. Near by is a solar food drier and a granary where food can be completely sealed from insects and rodents, and there are also community size versions.

The water lifting and storage demonstration includes two windmills, one locally designed, for raising water from shallow and deep wells, and a low-cost hydram pump made by a local Village Polytechnic, along with a variety of low-cost hand pumps. Water storage vessels include cement jars that can be easily made, in sizes from 300 to 3,000 litres (66 to 660 gallons).

There is a small garden to show the value of intercropping and proper composting, and the feed processing section covers a wide range of shellers and grinders, and a hand-operated oil press. Ways of making excellent weaning foods are among the other things shown.

This centre has become widely recognized as a place for the sharing and exchange of ideas, and since it was started UNICEF has been asked to help in setting up similar units, for agriculture extension workers and women's organizations elsewhere in Kenya; and also for teacher training, extension workers and government officers in several other African countries, including Lesotho, Swaziland, Ethiopia and Ghana.

The Village Technology Unit works closely with the Village Polytechnic Programme. This was started nearly ten years ago by the National Christian Council of Kenya, and is now run in remote areas by the Kenyan government.

The Village Polytechnics were started because of a growing concern about the number of unemployed school leavers: young people who have completed primary education, but cannot continue their education or find jobs. There were about 100,000 of them in 1970; there could be 300,000 in 1980. Ninety per cent live in the rural areas. The aim of the Village Polytechnics programme is to train primary school leavers in rural areas for work within their own community.

A Village Polytechnic generally has a local sponsor, which may be a church, a county council or other local body. When a local management committee for the Polytechnic has been elected, their first task is to make a survey of their community to identify the kind of job opportunities that exist and for which training is needed. The approach is far removed from conventional education and is strictly practical:

When making a curriculum you should keep in mind that to teach one skill you often have to teach another at the same time. If you are training a tailor for self-employment, he must also get some knowledge of book-keeping. If you train masons, you might also give them some basic skills in estimating building costs.

A V.P. curriculum must be flexible. The basic courses could be of two years' duration, but that must not keep you from launching additional shorter courses, like bee-keeping and vegetable growing.

And to have a course does not mean you have to have it for ever. Some courses could be run only once, others for a couple of years, and still others might be run only now and then. The yearly survey ought to give you good ideas on courses you should close down and new ones which you could start.

The basic rule is that you are training local people for *rural jobs* to keep money circulating in *rural areas*. Skills should be developed to stop money leaving rural areas. The farmer, who has been taught to make more money from his shamba, should buy his bread at a *local* bakery, his clothes from a *local* tailor, and improve his house using *local* builders. The V.P. should develop a community of skilled people who rely on each other.[9]

There are now well over 200 Village Polytechnics in Kenya, and their numbers steadily increase. They are mostly housed in very simple buildings; many have been built with voluntary help from their communities. The sponsoring authorities of the programme (the government and the N.C.C.K.) offer initial help in the form of field officers who will advise a community group wishing to start a Village Polytechnic. But financial assistance, tools and staff are provided only after a Polytechnic has been started and the community is supporting it with cash or in kind. Each Polytechnic develops its own programme related to local needs; but typical subjects include carpentry, masonry, tailoring and dressmaking, typing, baking, plumbing, blacksmithy and metal working. The Polytechnics try to keep track of their students after they leave, and so far it seems that over 70 per cent get work, or go on for further training, in their own localities.

Nigeria

Another highly successful production and training programme, firmly based on appropriate technology, is being run by the Relevant Technology Project in Jos, the capital of Plateau State in northern Nigeria. This was started in 1974, and is run jointly by the State government and the Bernard van Leer Foundation of the Netherlands.[10] The Foundation have supplied a single Technical Adviser to develop the project. From 1974 until his death in 1978 this was S. W. Eaves.

The objectives of the Project are to develop prototype products which can be manufactured locally and for which there is a local demand, and to train young people – primary school leavers and secondary school dropouts – in the process. The aim is to equip them either for work in local industry, or for self-employment in the locality. The training centres are co-educational and a number of places are also reserved for the physically handicapped of all ages who are trained to make (and sell) educational aids and toys.

The project now comprises four training centres, with a total staff of nineteen and an enrolment of nearly 200. Three of the centres are in Plateau State, in Jos, Pankshin and Langstang, and the others are in Benue State and Kano State. There are also five independently run workshops within the ambit of the project; each is owned by a group of up to five partners and employs as apprentices other young people. These workshops are located in villages and small towns in Plateau State.

The training centres are equipped with welding equipment, one or two locally constructed wood-turning lathes, and a variety of hand tools. The products made during the training exercises include:

agricultural equipment such as hoes, diggers, hand-operated threshing machines, cassava grinders, farm carts, and chicken cages and feeders;

medical equipment, including self-propelled wheelchairs for the disabled, hospital beds, delivery tables, bicycle-drawn ambulances, and crutches;

basic domestic furniture;

school furniture (desk, chairs, cupboards) and learning aids: numbers and letters sawn from plywood, educational puzzles, toys and geometrical forms.

Most of the products are sold and the proceeds are used to buy materials and supplies.

The training courses, which last a year, have several unusual features. One is that the centres have no fixed list of articles that the trainees learn to make during their course. Suggestions about what should be made may come from the staff, the trainees, or in response to an outside request. The staff then discuss whether there is a demand for the article, and whether it is suitable as a training exercise. If the proposal is accepted, a production cost estimate is drawn up, and a group of seven to ten trainees is assigned to making a prototype. If it proves successful, other groups of trainees go through the same processes of costing and construction. They also learn some elementary book-keeping.

Another characteristic of the centres is that all learning is by doing. Practical work is done as part of a project, the construction of a piece of equipment or furniture: the skills are learnt by making useful things. The trainees generally work in groups of five to ten, each working on the job in turn. The emphasis is always to show that what matters is not individual excellence but the total output of the team. Throughout the course, too, the trainees are given assignments which are described only in general terms, so that they have to find their own solutions to technical problems. This kind of combination of teamwork and individual initiative produces craftsmen who are educated for citizenship as well as craftsmanship – who are self-reliant, but at the same time concerned about their fellow workers.

Of those who have graduated from the centres so far, the whereabouts of ninety-two are known, and all are in employment. Thirty of them have set up their own workshops and between them employ between fifteen and twenty young people. This is very impressive, because all that the trainees get on leaving the centres is a set of basic tools and equipment to help them get started. They also receive advisory services from the project once they are on the job.

Even if this kind of training, or rather education and training, were to cost more, per trainee, than conventional forms of education, it would be a good investment. But in fact it costs far less. At the Relevant Technology centres the cost per graduate, including boarding, is about three-quarters of the annual recurrent cost per student in general secondary education. This must surely be the way to provide primary school leavers with a really worthwhile alternative to secondary education.

Tanzania

Tanzania is one of the poorest countries in Africa. It is also one of the very few countries, in Africa or anywhere else, whose development objectives are informed by a vision of economic and political decentralization, equality, and local and national self-reliance. It is taking longer, and proving more difficult, than many people hoped it would for Tanzania to build up the political and administrative structures to implement these ideas. The fact remains that it is probably the first country, outside China, to have identified, and asked itself the crucial question: what kind of mix of large and small industries is needed to create an industrial base, and at the same time create self-reliance through new skills and incomes at village level?

Rural development policy in Tanzania is based on the *ujamaa* village. The word *ujamaa* means 'the unity that exists between brothers', and it implies fairness, equality and common purpose. The great majority of families in such villages are very poor, subsistence farmers with only small amounts of cash income. In both the cases described below, the aim was to develop and introduce technologies that could be made and used by such families.

In the early 1970s a project was started by the I.L.O. and I.T.D.G., in collaboration with TAMTU (the Tanzanian Agricultural Machinery Testing Unit) to design and make prototype equipment suitable for village use. The first step was to discover what tasks (using what equipment) the villagers themselves regarded as the technical skills in the villages. 'In most villages there are men skilled in the use of the axe, the adze, the panga and the hoe. In many villages there are carpenters who make chairs, tables, beds, doors and houses using these implements, while in some there are smiths who forge adzes, hoes, knives and other small tools. This provides a reservoir of basic (mostly woodworking) skills which could be tapped.'[11]

It should be mentioned that TAMTU had a well equipped workshop making a wide range of agricultural tools and equipment, mostly animal-drawn, bought by relatively rich farmers in the Arusha region. Compared with conventional mechanized equipment, this is very low-cost: on average it costs about one-twentieth as much. But even this equipment was beyond the reach of the majority of very small farmers; they could neither afford the initial cash outlay, nor (as the equipment was made chiefly of metal) did they have the welding and other kit needed to maintain it.

The aim of the project was to develop a range of tools and equipment that could be made and repaired by the villagers themselves. A workshop was set up and equipped with simple tools at TAMTU, and after making several prototypes, similar workshops were set up in surrounding villages. Within a year, villagers were producing oxcarts, handcarts, cultivators, harrows, wheelbarrows and maize shellers. As people's skills grew, they also started making furniture.

The project maintained a very detailed record of costs. On average, the village level technology cost one-third as much as the machines offered for sale by TAMTU or by commercial manufacturers who are based in the towns. Of one of the machines, the cultivator, the head of the project has this to say:

Labour time for weeding is one of the notorious bottlenecks in peasant agriculture and lack of adequate weeding can account for heavy losses in yields. Weeding between rows is traditionally carried out by hoe and requires much arduous labour. But it is technically quite feasible to do the weeding by an animal-powered implement called a cultivator. This machine is pulled along between the rows of plants with its adjustable blades cutting the weed roots. Hand-weeding can then cope with the few remaining weeds between plants. At 192 (Tanzanian) shillings, however, the metal cultivator of Ubungo Farm Implements (the factory in Dar Es Salaam), although an admirable device, is costly for the subsistence farmer. Breakdowns occur occasionally, and repairs or spare parts require a cash outlay at weeding time which is several months after the harvest and therefore a time of cash shortage. The project therefore designed a cultivator which could be made and repaired in the village. The frame was made of trimmed poles or branches bolted together, and with simple depth and width controls. The tines or blades were made from old car springs, and could be shaped in the village using simple metal working techniques of heating, hammering and cutting. The cultivator was designed to have a low resistance so that one donkey or ox could pull it. The implement was well received by farmers who were soon able to make their own versions for special needs. At a unit cost of just over 50 shillings, with a cash outlay of just under 29 shillings, it represents an appreciable saving (compared with the 192 shillings for the factory machine).[12]

Although the difference in total cost between hand weeding and mechanical weeding is not great, the use of this simple piece of capital equipment greatly reduces the labour required: and given the labour bottleneck on weeding, this is a gain of great significance to the farmer.

The same was not true of all equipment. Thus although it was possible to make a plough at village level, it cost much the same but was inferior to the iron plough made in a more conventionally equipped workshop. Farmers were quick to see this, and also to understand that they could afford the cash to buy a good plough if they got other implements, which could be made more cheaply, in the village.[13]

This, then, was one of the first practical, carefully documented demonstrations of the fact that villagers, with the help of a simple and inexpensive kit of tools, could make and maintain a wide variety of farm tools and equipment – and also make many of the tools needed for the purpose. By substituting wood for metal wherever possible, and using materials which could be obtained for a non-cash labour 'cost', it was possible to make not only the farm equipment mentioned earlier, but also production tools such as a workbench, vice, mallet, bellows, chisel, adze, axe, bradawl, screwdriver, knife, hammer and pliers. This is, of course, the first step towards self-reliance.

The process of matching technologies with the real needs of *ujamaa* villagers has been further extended by the more recently established Arusha Appropriate Technology Project. This is a small voluntary organization which works in close collaboration with the regional development officers and with village people. To begin with, they involve the village in identifying its own needs and priorities. Village committees that are interested in innovation are asked to form a survey committee of three men and three women. The A.A.T.P. show this team how to carry out a survey, and after meetings with individuals and groups they put their suggestions forward for approval by the village committee and at a general meeting.

One of these survey teams, which had identified water supplies as the highest priority, also said that they would like to get it by means of a hand pump. This is the second stage of the A.A.T.P.'s work, that of matching a technology to a particular need.

In this case they designed a hand pump that could be built from parts easily available locally, using simple hand tools which do not require electric power or welding. After a pump had been made and fully tested in the village, people from surrounding villages were trained and they decided to start a small production co-operative. Orders for pumps are now coming to them from the Water Ministry.

The A.A.T.P. has also successfully introduced methane gas production and windmills for water pumping to villages in the region.

The standard Indian design of gobar (methane) gas generator, a seventy cubic foot metal tank, costs over £800, which is far too expensive for Tanzanian villagers. After trying out several alternatives, A.A.T.P. finally came up with a similar unit costing about £130, and a number of local co-operatives have started to make and sell these. But even this is beyond the reach of the majority of local people – most of whom have several cattle but little cash. The latest model comprises seven oil drums grouped together in an open pit; this costs less than £50. It is too early to say whether this version of the methane plant will catch on, but even now the A.A.T.P. have demonstrated and proved a method of cooking that will prove to be of increasing value to African countries, especially where wood is becoming scarcer but cattle are relatively plentiful.

Windmills are not a poor family's technology, nor are they a village technology: but they can be made in small urban workshops, and they are within the reach of many rural communities. A.A.T.P.'s windmill programme started with the erection of a windmill at the A.A.T.P. workshop, and another at Arusha airport which pumps water up 200 feet. A co-operative has now been set up near A.A.T.P. which can make one standard windmill a week, costing about £1,100 including the pump. This is about half the local cost of a diesel pump, and about one-sixth that of the cheapest imported windmill. A.A.T.P.'s plans for extending its windmill programme include setting up two field teams to erect the windmills and train local technicians in maintenance and running repairs. The 'service centre' will be the co-operative which makes the windmills and it will undertake major repairs and overhauls.

The A.A.T.P. is an excellent example of how to bring the right kind of technology to the right people. It starts by involving the villagers in specifying their most urgent needs, and its end objective is to put both the production and the use of the technology into the hands of village people.

Latin America

Organizations dealing specifically with the development and dissemination of appropriate technologies are much thinner on the ground in Latin America than they are in Africa or the Indian sub-continent; neither is there, as yet, the same growth of interest in the subject on the part of governments.

There are, of course, exceptions. Four examples are given below. Two are large and well funded organizations which tend to concentrate perhaps too much on technologies and lose sight of people. The other two start with people, and arrive at appropriate technologies.

There is, to begin with, the integrated development effort of the Las Gaviotas programme to establish ecologically sound settlements in the Orinoco basin of Colombia. This includes a major appropriate technology component operated by the technical department of the University of Los Andes in Bogotá. As part of this project, workshops are being set up to manufacture a variety of appropriate technology machines and equipment, prototypes of which have been installed for field trials. A solar heating panel provides hot water for the hospital. A cassava grinder is in use, a micro-turbine and generator are operating on a low head of water produced by damming a stream, and the same head works a hydraulic ram pump. Windmills raise water from two wells. After these items have been field tested and modified, it is intended to set up workshops to produce them in large quantities.

In this case, although the technical work is reported to be of very high quality, it remains to be seen whether many of the 650,000 or so poor *campesinos* will be able to afford such equipment, even on very easy credit terms.[14]

Another very ambitious A.T. programme is that recently started by SENA, the public corporation that runs all the professional training of skilled labour in Colombia. SENA now has a professional teaching staff of between 7,000 and 8,000 and a budget of about U.S. $70 million. In 1976 SENA decided to build appropriate technology into its training programme, with special emphasis on the traditional rural sector, artisans, and small enterprises. Today the task of identifying appropriate technologies, thoroughly field testing them, and incorporating them into training programmes is the prime task of one of SENA's central divisions (the Subdirección General Tecnico Pedagogica).

In the small-scale sector it operates Urban and Rural Mobile Programmes, which send out technical instructors to small urban firms, and to small farmers and rural groups, to diffuse the best existing technologies and introduce proven and tested innovations. Their contacts with small industry over the years – they have assisted over 6,000 small enterprises – enable SENA to draw on a large reservoir of data on existing and new technologies. But in the rural areas SENA has found that formal training methods do not work; so it is now in the process of learning how to identify the constraints – technological, economic and organizational – on small farmers' productivity, and how to remove these constraints with the collaboration of people in the rural areas – teaching people how to identify and solve their own problems. One of its most promising development projects is that of helping small farmers to set up small units for processing tropical fruit. It is also working on fish-farming, and small farm mechanization.

SENA is held in very high esteem by the Colombian people, and if it can modify its rather formal training methods to meet the real needs of poor farmers and artisans, SENA could become a major force in rural and small-scale development.

Two other Latin American A.T. organizations that are much smaller, more flexible, and certainly very effective, are Futuro para la Niñez (Future for Children) in Colombia, and CEMAT (Centro Mesoamericano de Estudios Sobre Tecnologia Apropriada) in Guatemala.

Futuro is a small voluntary organization operating in eight urban settlements and about 150 rural villages. Communities are encouraged to provide a better world for their children, by working together to improve living conditions, to provide a more varied and healthier diet, and to provide for their children's welfare. Futuro's staff (five full-time and two part-time) encourage *campesinos* to establish their own priorities and nothing is done which they cannot do for themselves. Futuro has thus promoted schemes of sewage treatment; composting; terracing of steep hillsides; water pumping; irrigation; fish culture; tile making; and cassava and fibre processing.

Of these activities, the terracing of hillsides for vegetable growing, and the building of fish ponds, are the most popular; both are helping to generate income as well as to raise standards of nutrition. A typical terrace measures one and a half yards by three yards and is filled with organic compost; the only cash outlay is for seed. The average family makes seven terraces, each of which can produce three vegetable crops a year. At local prices, the family can thus consume or sell about U.S. $50 worth of vegetables a year. Similarly, fish ponds require no cash outlay, and the fish are fed on grass and kitchen scraps. From a typical pond a family might consume between U.S. $40 and $50 worth of fish a year. The Futuro team observe that the idea for the vegetable terraces and fish farms came from listening to the views of small farmers. In the region around Medellin, Futuro's base, many thousands of people are discovering how to turn their labour into something useful.[15]

Guatemala City is the headquarters of CEMAT, a non-profit organization formed in 1976 after the disastrous Guatemalan earthquake. It has a permanent staff of nineteen working on housing, health, energy, agriculture and education, and building up an information service of technical literature on A.T.

Five permanent consultants work on CEMAT's housing programme. An inventory has been made of housing types, occupation, construction materials, and so on, and CEMAT organizes seminar workshops to promote the better use of locally available materials such as adobe and terracreto (volcanic soil and lime). A peasant-operated pilot cement plant is projected for the earthquake zone.

Work on health includes the training of more than twenty rural health auxiliaries, the documentation and promotion of local medicinal herbs, and the development of latrines linked with biogas production.

CEMAT's most extensive programme is currently in the field of energy: the diffusion, on a self-sustaining basis, of an imaginative but simple technology in the form of a domestic clay stove which reduces the consumption of firewood by up to 50 per cent.

As in many other countries, deforestation for firewood in Guatemala contributes to soil erosion and drainage problems; and the growing shortage of firewood means that peasants have to travel greater distances to collect it, or pay higher prices to buy it. Hence a stove that saves between one-third and one half the present firewood consumption would be both a domestic boon and a national asset; that is what the Lorena stove can do.

The Lorena stove was developed from an Asian (Indian) design, and thoroughly tested. Before launching its present programme CEMAT first trained representatives of the Save the Children Federation in Honduras in the Lorena technology, which led to its dissemination there. This is only one of the Federation's many activities in the field of appropriate technology.

The stove is cheap; it costs between $15 and $20 to build, and this cost is recovered by savings on firewood within two or three months. Other advantages are that its chimney keeps the kitchen free of smoke; it is free standing and provides a raised flat surface on which to cook; it cuts the time needed to prepare meals by at least one-third; it has three burners which can provide various intensities of heat; and it can be made of locally available materials. Most important of all, the extensive field tests showed that people like using it.

CEMAT is now running courses in the homes of villagers in Guatemala's highlands. At these, groups of ten to twelve people are being taught how to build and use the stove. Experience based on trial runs suggests that about half of each group will opt to build stoves in their own homes, and two or three will take up building the stoves for others in their spare time. Since half the cost of a Lorena is labour, this cadre of craftsmen can earn between $8 and $15 per stove.

An estimated 400 to 500 stoves will be built during the two years that this programme will run, and thereafter the technology is likely to take off on a commercial basis in the hands of local craftsmen. A technology more attuned to the needs and resources of poor people would be hard to find.

Supplement B

FURTHER ALTERNATIVE ACTIVITIES IN BRITAIN

There is now a considerable body of organizations providing supporting services for local enterprises of all kinds. Many of these are briefly described below. Broadly considered, the first three headings – local authorities, big firms and voluntary agencies – relate to support for conventional small companies and co-operatives. The second group comprises organizations that are working on less conventional lines, for example, the Centre for Alternative Technology, New Age Access, Eco 2000, Pocklington School, the Universities of Reading and Warwick, the Open University, the Natural Energy Centre, and local work on low-energy housing.

Then there are groups that combine advocacy with local action, notably the Foundation for Alternatives, and Friends of the Earth; and information and pressure groups concerned with the environment, food and energy, alternative medicine and health care, and town planning. Finally there is a brief note on the literature supporting the alternatives movement in Britain.

Besides the local, community-based schemes to help small enterprises start up, and the central government's somewhat sketchy programme of assistance – DoI Small Firms Information Centres, COSIRA, Training Services Agency 'How to Start Your Own Business' courses at the Durham, London and Manchester Business Schools and the like – many local authorities, large companies, voluntary organizations, universities and the B.B.C. have taken independent steps to assist small firms to get started. Most of these initiatives have been concerned to foster local enterprise in inner city areas but they all have the common aim of supporting and making business life more acceptable to the small-scale entrepreneur. There is only space here to give a few examples of the quality and type of ventures which have been started.

Local Authorities

In Scotland the Industrial Development Unit of the Strathclyde Regional Council has begun a scheme for New Enterprise Community Workshops. These provide shared workshop facilities, technical advice and guidance on establishing and running a business for local people who have product ideas, or new process inventions they would like to develop to the point where they can be commercially tested. The two workshops at Hamilton and Paisley have proved very successful – demand for space in them has been consistently high and the scheme is likely to be extended to other parts of the region. The Scottish Development Agency has taken a constructive interest in the scheme and for many years has run its own Small Business Division which recently began to recruit retired business executives to give specialized help for small firms and first-time entrepreneurs.

Other local authorities in Britain are beginning to recognize the importance of attracting small firms to their areas and have appointed Industrial Development Officers with a brief for small enterprises. Some have even financed campaigns to make grants or loans to small companies, such as the Merseyside County Council Help Active Small Enterprise (CHASE) scheme, which has granted £25,000 to the St Helens Community Trust (see pp. 000–00). In Lancaster, the City Council runs a flatted workshops project for scientific and design-based small firms. These also have access to apparatus and technical advice from Lancaster University. In London, the North Southwark Project is an advisory and consultancy scheme run jointly by local businesses and the local authority with two people seconded from I.B.M. Ltd to make available managerial and specialist services to local small firms.

Big Business Helping Small

The support that large companies and industrial and professional bodies have been giving to small enterprises has little altruistic quality about it. Large firms have recognized that their business and financial interests and the social context in which they are judged can be improved by taking measures to facilitate an increase in the number of small firms.[1]

The British Institute of Management and the Confederation of British Industry run projects to assist and advise small firms and, in May 1979, under the auspices of the London Chamber of Commerce, the London Enterprise Agency was launched. This project is financed by nine major companies and is working to set up flatted workshops for small firms in the docklands, Wandsworth, West Norwood, Hackney and other districts, and provide consultancy and advisory services to newly started and established small companies. Enterprise agencies have been suggested for a number of other major cities in the U.K.

Several of the companies involved in the London Enterprise Agency also run independent schemes to assist small enterprise – I.C.I. Ltd, Marks and Spencer, I.B.M. Ltd, and Shell (U.K.) Ltd. At the same time, they have set an example being followed by other large commercial organizations and branches of central and local government in recasting their purchasing, contracting and credit supply policies to favour small firms.

Voluntary Agencies

Voluntary agencies, working at both local and national levels, have begun various initiatives to foster small enterprise. The Action Resource Centre has a unit in Islington, London, counselling small firms faced with relocation under the local authority's redevelopment scheme and provides many other organizations with experienced personnel seconded from large companies. The Liverpool Council for Voluntary Service (C.V.S.) which, over the years, has run numerous research and action projects on unemployment in central Liverpool, has employed A.R.C. through a £40,000 grant from the Rowntree Memorial Trust to organize a programme of secondments from industry to local projects offering new, long-term work opportunities. Other C.V.S. branches have been just as active in projects to combat local unemployment, especially among young people and ethnic groups, and to open up new work opportunities – in Leeds and Haringey, London, for example, they have been actively promoting the formation of local Co-operative Development Agencies. In Glasgow the C.V.S. has been instrumental in setting up the 'Goodwill' furniture refurbishing workshop and store, a self-supporting business with an annual turnover in excess of £50,000, which employs fourteen people.

In London, URBED (Urban and Economic Development), an independent, non-profit making organization, has been working closely with inner city communities – Covent Garden, Hackney, Southwark and Tower Hamlets – on ways to regenerate local communities and re-establish a thriving economic base, particularly through the encouragement of small business. URBED, funded by the Gatsby Foundation, is able to draw upon a wide range of professional consultants for research and promotional activities.

Other initiatives run by voluntary bodies include Enterprise North, which organizes seven panels composed of local businessmen from the north-east to supply new and established small firms with advice on product viability, finance, marketing and so on; and Northumbria Habitat, a registered charity based in Durham that undertakes market surveys, trade exhibitions, assessment of premises and other practical research for small businesses. This group emerged from the energetic activities of Friends of the Earth, Durham.

In Strathclyde, the Local Enterprise Advisory Project (LEAP) was set up in 1978 to assist community groups in three areas of high unemployment: Ferguslie Park, Paisley; Govan, Glasgow; and East Greenock, to investigate and prepare schemes to provide long-term viable employment. LEAP is funded for a two-year experimental period under the Urban Aid Programme and employs a full-time organizer, John Pearce. The LEAP experiment is an extension of the community development principle to industrial development.[2] By providing resources from the Local Government Research Unit of the Paisley College of Technology and local community resource centres such as the Govan Area Resource Centre, which can make small grants to community groups, it is hoped that active field work, information and finance for feasibility studies will encourage local enterprise. The project has also produced educational materials, a video cassette, 'Now Make Your Own Jobs', and run a local enterprise competition. Local response so far has been good, and John Pearce is assisting a group in Govan, through the Employment Study Group of G.A.R.C., to set up a co-operatively run shop, and assessing the feasibility of a workspace project for small businesses. He is also working with the W.E.A. and trades unions to publicize the resources LEAP offers. In the other two areas the project covers, work is progressing more slowly but there is a growing confidence in all three communities that they have the ability to re-create employment in areas that have been deserted by large-scale enterprises.

Two further examples of local groups seeking to create new employment are the North Wales Employment Resource and Advice Centre and Rhondda Enterprises. The north Wales group employs two full-time workers on local enterprise advice and information campaigns backed by the Urban Aid Programme. Rhondda Enterprises in south Wales is an M.S.C.-financed Training Workshop which also runs a workspace scheme for first-time entrepreneurs. It was started with the co-operation of local trades unions, businessmen and the local authority. One of their aims is to start small co-operative production units using the workshop as a base; they also have plans for local horticulture.

The Institute of Cultural Affairs (I.C.A.) promotes community action on an international scale. The I.C.A., an associate group of the Ecumenical Institute, is a world-wide organization with its headquarters in Chicago and offices in Brussels, Singapore, Bombay, Hong Kong and Nairobi, and over 100 other major cities serving thirty-two nations. The Institute aims to stimulate community action and local development by encouraging each community to become conscious of its own unique worth and hence begin confronting the problems which beset it. Voluntary consultants and specialist advisers from all over the world are available to offer their advice and bring a fresh perspective to the local problems of each area, which are carefully analysed. The I.C.A. does not control the nature or direction of changes which each community expresses a desire to see implemented, but acts as a catalyst and forum for local discussion, analysis and formal recommendations. The Institute has been active in the U.K. since 1971 and has been gradually expanding its work of training local people in community development methods over the past eight years. Its main work in Britain is its Human Development Projects on the Isle of Dogs, London, and at Tai'gwaith, a former mining village in Wales.

A.T. Demonstration Centres

Centre for Alternative Technology (CAT)

CAT was set up in a disused slate quarry, overlooking the Snowdonia National Park in 1974. It is now the largest and best known alternative technology demonstration and research centre in Britain.

The site is independent of all mains services and supports a full-time staff of twenty people, under the Director, Roderick James, who carry out basic research and monitoring of the A.T. equipment installed in the Centre. This includes thirty commercial solar and D.I.Y. panels for hot water, electricity and tracking parabolic reflectors; a dozen varieties of windmills and aerogenerators for electricity and pumping; a water turbine and timber D.I.Y. water wheel, and other projects, such as waste recycling, heat exchangers and methane production. The site includes a conservation house with 18 inches of wall insulation, heat pumps and quadruple glazing, and a solar-heated exhibition hall.

The purpose of the Centre is to demonstrate and develop A.T. techniques and horticultural practices that are compatible with sustainable and ecologically desirable lifestyles. They run a small-scale engineering workshop based on a blacksmith's forge; a small-holding with goats, cows, ducks, rabbits and a methane plant; various sizes of vegetable plots; an organically cultivated allotment and a fish farming pond.

Visitors to the Centre, open between February and November, number about 50,000 a year. There is a well stocked bookshop with a wide range of books, pamphlets and journals on D.I.Y., crafts and A.T. Many of the publications are by the Centre's staff, such as D.I.Y. plans for a 5 watt wind generator, pumping windmill, timber water wheel and over twenty information sheets on energy production, organic gardening and farming, and waste recycling. CAT has also published a stimulating account of the possible non-nuclear, renewable energy options open to the U.K. up to the year 2025.[3] A newsletter is published twice a year and gives full coverage of all the new work and events taking place at the quarry site. CAT has also produced a 'Useful Addresses' sheet listing over 100 contacts in alternative energy and medicine, crafts, environmental groups, co-operatives, recycling, organic gardening and farming, and wholefoods.

In nearby Machynlleth, CAT has recently bought a ten-roomed shop which is being converted to house a wholefood store, crafts department and possibly a restaurant and bicycle renting service. At the quarry future plans include a 300-seat lecture hall, which is now under construction, and a schools programme on smallholding.

CAT is sponsored by the Society for Environmental Improvement, a registered charity, and has been built up largely through generous material and financial contributions from over 130 U.K. firms, and a great deal of voluntary labour.

New Age Access (N.A.A.)

This educational charity based in Hexham in rural Northumberland has been active since 1974 in A.T. research on rural lifestyle and environmental education projects.

N.A.A. was set up by Bryan Dale, Geoff Watson and Phil Smith and now comprises nine full and part-time workers. There are about 150 Friends of New Age Access, who can be called upon to provide extra help for large projects such as festivals or outside exhibitions.

The educational function of the group has naturally been an important aspect of their work, and they have organized a great number of talks, practical demonstrations and workshops at schools and colleges in the U.K. N.A.A. also handles a wide variety of requests for advice on other A.T. and environmental groups, technical details of N.A.A. projects and research and appeals for exhibition material. Their mobile exhibition van has been in great demand. In 1978 N.A.A. dealt with technical enquiries from as far afield as Helsinki, Scotland and South Korea, as well as giving detailed practical assistance to Byker City Farm in Newcastle, Friends of the Earth, Durham, and many other local organizations.

The types of requests for advice that have been coming in to

N.A.A. since 1974 have obliged the group to undertake research and field work for subjects where practical information was limited or difficult to obtain. They have produced a very useful educational package, *Plan-it*, containing information sheets on organic gardening, solar and wind power, and nutrition; and D.I.Y. plans for the N.A.A.-designed beehive.

Their research on renewable energy sources has yielded a widely acclaimed design for a vertical axis wind turbine, the Maximill. This was developed and tested in conjunction with a joint project team from the Institutions of Mechanical Engineers and Electrical Engineers and built and tested at Newcastle University.[4] N.A.A. are currently developing a D.I.Y. Minimill with the assistance of local businesses. The high quality of research and technical competence that the group have displayed in their work on wind energy has made them a well respected authority on the subject. Their handbook, *Windworker*, is an expertly written, comprehensive guide for the layman on all practical aspects of wind power. N.A.A. have designed a D.I.Y. solar collector which is part of a travelling exhibition display. Further designs for solar collectors and other renewable energy sources are being investigated.

To enable them to expand their R. and D. programme on organic horticulture, animal husbandry, small-scale building systems and rural crafts, N.A.A. are raising funds to buy the thirty-acre self-contained site, buildings and gardens of a former isolation hospital near Hexham. The site is planned as a permanent base for educational projects, lectures, workshops and publications distribution. It will also be a public demonstration and advisory centre dealing with the techniques and practices for sustainable rural lifestyles, and will emphasize the self-help, self-reliance aspects of A.T. rather than simply display commercial hardware.

Also based in Hexham, in a former tannery, is an off-shoot company, Northumbrian Energy Workshop Ltd. This co-operative workshop does a thriving business manufacturing and selling wind power equipment and small-scale engineering plant. It undertakes as well industrial and domestic energy conservation schemes; provides legal consultancy services on co-operative ownership; and provides graphics and design services. This enterprising commercial venture is intended to be both a source of practical information, and of funds, for New Age Access.

Eco 2000

The original idea for an A.T. centre near Exeter in Devon came from the announcement of the government's Job Creation Programme in 1975, when Ken Penney saw an opportunity to use J.C.P. funds to employ redundant local craftsmen to teach their skills to unemployed youngsters through the construction of solar collectors, windmills and low-energy housing. Unfortunately, this proposal was turned down by the local J.C.P. unit.

Undaunted, Ken Penney pressed ahead with an idea for an A.T. exhibition centre that would provide facilities to employ young people to be trained in A.T. craftwork. A film about the aims and plans for this centre was broadcast by B.B.C. Plymouth and attracted a favourable response and many offers of support. The Devon County Council became keenly interested in the project, and with the enthusiastic assistance of their Architects' Department a feasibility study for the centre, to be run by Eco 2000, an educational charity, was prepared.

The study proposed buildings constructed of local stone and incorporating solar panels, a Trombe wall and extensive insulation; workshops and craft centres; and organic horticulture, waste recycling and animal husbandry. The scheme is simply and flexibly designed so that it can be readily put into practice by unskilled and semi-skilled labour. Employment creation is an integral part of the scheme, and the centre is envisaged as a local base for helping to start up community co-operative enterprises. Additionally, the site will include A.T. information and advice services, and educational services to local schools and colleges.

Although the original site for which the study was prepared is no longer available the central features of the proposal still stand, and these will be included on the alternative site which has now been promised for the Eco 2000 centre – part of the 144-acre Storer Park near Exeter, and adjacent to the A38, a major tourist route to Plymouth.

Educational Initiatives in A.T.

Within educational institutions in the U.K. very little attention has been given to the wider implications of appropriate technology.

Many universities and polytechnics have been fairly quick in setting up small research programmes on alternative energy sources. On energy use in buildings, for example, there are now more than sixty research projects at universities and government research establishments; twenty-five of these deal with district heating pumps and solar heating. Overall, however, the government has made only modest financial contributions to research into renewable energy sources, concentrating the bulk of the energy research grants on nuclear power, and what research is taking place is on very large-scale renewable sources of electricity generation. Universities and other centres of formal education have shown a general reluctance to extend their research beyond the brief flirtation with one or two solar panels or a windmill, and develop it further into the field of appropriate technology and small-scale, low-cost engineering design. There are, however, some notable exceptions.

Pocklington School

In 1974, John Jeffery and some of his colleagues at Pocklington School, an independent secondary school near York, started up a 'Possible Futures' course for their sixth form students. The course gave students an opportunity to examine in detail some of the technical and social aspects of appropriate technology. Lectures from specialist outside speakers were complemented by background studies of Third World problems, nuclear power and self-sufficiency.[5] Coursework was linked to the school design centre, which was set up in 1969 to provide facilities for pupils of all ages to work on a broad range of art, craft and technical activity including engineering projects that related directly to real problems in the 'outside world'.[6] As far as is known, this is the only attempt made by a secondary school to introduce practical and academic work in the spirit of A.T. into the conventional school syllabus.

Universities and Colleges

The number of universities, colleges and polytechnics offering training courses, research back-up and/or consultancy services to small enterprises is growing. There are now over fifteen such institutions already well established in this field and they cover nearly all parts of the country. The Durham University Business School (DUBS) for example, set up a Small Business Centre in 1971 in conjunction with twenty small firms. The Centre runs training courses, seminars, counselling and research programmes for small firms in the north-east. DUBS is also the base for the New Enterprise Development Project which helps to co-ordinate and support the efforts of Enterprise North by providing specialist advisory and education facilities for entrepreneurs. It is financed by the DoI, Rowntree Charitable Trust, I.C.I. Ltd, and Shell (U.K.) Ltd. The latter two firms have given further assistance to the project: I.C.I. has seconded an executive for three years and Shell has helped organize a national New Enterprise competition run by the project team.

The National Extension College ran a series of courses in 1980 entitled 'Your Own Business' based on their excellent Small Business Kit (which includes a special section on co-operatives). These courses coincided with the B.B.C. programmes on setting up and running a small business, and were based at various towns in the south-west of England.

The Department of Engineering, University of Reading, has long been associated with the Power Panel of I.T.D.G., and has built up an extensive programme for development and testing of low-cost, renewable energy systems; it is currently working on vertical axis windmills, water turbines, pistonless pumps, variable speed pumps, heat exchangers and solar pumps.

The University of Warwick has recently announced a new three-year B.Sc. degree course which has been available since October 1980 – Engineering Design and Appropriate Technology (EDAT). This is the only undergraduate course of its kind in the U.K. and had an intake of twenty students in its first year. EDAT provides a firm grounding in mechanical and production engineering in which economic, cultural and environmental considerations are given prime importance. Agricultural technology is singled out for particular attention. It is primarily a vocational course for engineers who intend to work in small firms or co-operative groups, on rural development overseas or with self-sufficient communities in Britain.

The Open University at Milton Keynes is the home of the Alternative Technology Group (A.T.G.) formed in 1976 to advance and disseminate the knowledge and application of ecologically sound techniques. The group has been particularly concerned with developing technical knowledge relevant to the needs of decentralized, substantially self-sufficient and economically self-organized communities, and with identifying ways in which such communities can evolve within industrialized societies. Its work in pursuit of these aims has concentrated on assessing the practical feasibility of achieving various levels of self-reliance in basic necessities, for both the family unit and the small to medium size community.

The research programme currently includes projects on food, shelter, energy, transport, manufacturing and recycling, and policy studies relating to such topics as the job creation potential of alternative energy technologies, the potential of alternative corporate plans in large-scale industry and the co-ordination and feasibility of community-scaled decentralized production of food, energy and manufactured goods.

The Group has published a splendid series of working papers and other publications which provide a valuable record of their work and individual research interests. Various courses produced by the Open University Faculty of Technology include A.T. elements: *T-262 Man Made Futures: Design and Technology*, which includes files on The Future of Food, and Shelter and Work; and *T-361 Control of Technology*, which analyses political and community strategies for choosing and implementing appropriate forms of technology – Godfrey Boyles's Unit (10/11), 'Community Technology', and Alan Thomas's and Martin Lockett's Unit (14/15) on Appropriate Technology are particularly useful.[7] These courses are backed up by television and radio programmes including on the spot interviews with people who are involved in the day to day running and organization of working alternatives in housing, local planning, food production, technology, and so on.

The A.T.G. has also been very active in the local community. It has designed and built a prototype power assisted tricycle, the 'trike', equipped with two children's seats and a shopping platform and powered by a motorcycle battery and small electric motor for use in the undulating and widely spread site of Milton Keynes new town. Other projects include a small-scale paper recyling plant for the town, wind power, heat-conserving greenhouses and insulation. In all cases the A.T.G. places great emphasis on devising small-scale, co-operative means of production.

An integral part of the research activity of the A.T.G. has been 'to create and study situations in which the related activities of self-sufficiency and alternative production can be put into practice', and the Group have been working very closely with four projects, which they have helped set up, to investigate the practical implications of A.T. lifestyles.[8] (As these projects are in family homes, visits can be arranged only through the A.T.G.)

DOMESTIC-SCALE SELF-SUFFICIENCY STUDIES: This work is being carried out by Robert and Brenda Vale on a 1¾-acre smallholding in East Anglia. (Robert and Brenda Vale formerly worked with the Cambridge University Antarctic Housing Project, and with the Biotechnic Research and Design – BRAD – commune in mid-Wales, for which they designed a very successful low-energy house with solar water heating.) Their main concern has been food and energy self-sufficiency. They have converted their cottage to low-energy demand through extensive insulation, and rely largely upon renewable energy sources – solar, wind power and wood. On the smallholding they practise highly productive organic gardening techniques combined with intensive livestock husbandry. By carefully monitoring all supplies (including labour) and production, their aim is to demonstrate how it would be possible for a uniform distribution of the U.K. population to feed itself from the available agricultural land. The work so far has yielded a substantial body of detailed practical information.[9]

THE REDFIELD COMMUNITY: started by Peter Read and others, this is based in a large Victorian house set in eighteen acres of mixed woodland, pasture and kitchen garden near Winslow in Buckinghamshire. The community is made up of about twenty adults and their children. Initially the project set out to define a legal and financial structure in which all the members could participate as equals, notwithstanding their individual financial contributions. This also led them to examine the institutional barriers to living as a collective – lending policies of building societies, local planning restrictions and the like.

Currently the project is setting out to become more self-sufficient in food and energy production. The amount of land available per head of the community is roughly the same as that available to the Vales in East Anglia, and this may allow comparative self-sufficiency studies to be carried out.

THE 'THIRD GARDEN CITY' PROJECT: this project arose out of a proposal made in 1978 by Lord Campbell, Chairman of Milton Keynes Development Corporation (M.K.D.C.) to the Town and Country Planning Association (T.C.P.A.) in London in which he stressed the 'astonishing relevance' of Howard's ideas today, and remarked that Howard's

> vocabulary included small scale settlements; a basically co-operative economy; a marriage of town and country; control by the community of its own development; control by the community of the land values it creates; the importance of a social environment in which the individual could develop his own ideas and manage his own affairs in co-operation with his neighbours; and the strength of the family unit in the community. And Howard wanted dispersal in order to make possible the humane redevelopment of the inner city.

Lord Campbell suggested that the T.C.P.A. should draw up proposals for a 'Third Garden City', following the now famous garden cities it established at Letchworth and Welwyn early in this century. This new settlement should incorporate not only the original garden city ideals, such as co-operation and participation, but also the more recent ecological imperatives of energy and resource conservation. M.K.D.C., he suggested, might be willing to make available some land within Milton Keynes for such an experimental community.

Since Lord Campbell's proposal, and the publication in 1979 of the T.C.P.A.'s *Prospectus for a Third Garden City*, the A.T. Group has become intimately involved in the detailed negotiations which have taken place between the T.C.P.A. and M.K.D.C., mainly in its role as consultant to the Greentown Group, a local pressure group consisting of the potential residents of the new settlement. In mid-1980 the M.K.D.C. approved (and a Rowntree trust funded) a pilot project, now becoming known as 'Greentown', in 'Garden Village' form on a site of twenty-five to fifty acres with several hundred inhabitants, to go ahead in 1981–2.

THE RAINBOW HOUSING CO-OPERATIVE: this project was started by Godfrey Boyle and Martin Lockett of A.T.G., with others, who helped to organize the renovation and extensive insulation of twenty-four railwaymen's cottages by the M.K.D.C. These have been leased to a Tenants' Housing Co-operative since June 1978. Various problems were encountered, not least the difficulty of motivating and negotiating with a public body about how to get it all going. But now that the co-operative has been operating successfully for more than two years, initial scepticism has been replaced by considerable enthusiasm for housing co-operatives, and further co-operatives are in the process of being formed.

The amount of spare land available, one acre, is too small to sustain complete self-sufficiency in food, but the co-operative does grow a large proportion of its own vegetables. In addition, several members of the co-operative are now working either full-time or part-time at the nearby St James Street Workshops – a project also inspired by the A.T.G., which includes ten small co-operatively run light industrial workshop units housed in a converted school.

The Foundation for Alternatives

There is a very wide range of organizations and individuals advocating alternatives, providing information about who is doing what, where, and generally creating a climate of opinion favourable to the promotion of what James Robertson has aptly described as 'sane, humane and ecological' lifestyles, technologies and social and political institutions. Below we offer a brief survey of some of these groups and the fields of activity they cover.

For several years past the Foundation for Alternatives has been promoting models of decentralized, small-scale, community enterprises. Established in 1977 as a charitable company, the Foundation took over responsibility for the projects of the earlier Alternative Society. Its purpose is to mobilize the existing resources of the local community in whatever field it works and foster a co-operative endeavour which leads toward greater personal autonomy and local responsibility. It is essentially an enabling organization rather than an administrative body, and provides an initiating research and servicing centre for several relatively independent projects, but does not manage their activities.

The projects have a clearly defined function; they are 'feelers' put out into the future, ways of testing on a small scale new models of social organization. Therefore, they are characterized by certain hallmarks. In particular, they meet a socially relevant need, satisfy that need in a radically new way, monitor and record all their work, are capable of being reproduced, and, after the trial period, are followed up by appropriate publicity and political education.

By May 1979 there were nine projects associated with the Foundation in various stages of development. The work of one, the Association of Local Enterprise Trusts, has already been described in Chapter Four (pp. 121–2). This is being done in collaboration with the Intermediate Technology Development Group.

The other eight projects are:

Local Initiatives

The Foundation published at the beginning of 1980 its first set of thirty profiles of Local Initiatives in Great Britain.[10]

Local initiatives are self-help efforts by local communities which are waking up to their own responsibility for the locality – not just in social matters, but also in the management of the local economy, and the use of local resources. They may be Local Enterprise Trusts, Community Co-ops, or simply informal groupings of local citizens. There is no stereotype.

The growth of such initiatives, which is now happening in many different countries, is a very hopeful sign for the future. Releasing the energy of small communities is like splitting the atom. It is hard to do; but once it is done, the results in terms of energy release are incalculable, and we need all the energy and initiative we can get to solve the problems that lie ahead.

The Foundation for Alternatives has decided that its general policy during the next five years will be to serve the Local Initiatives movement. In place of the present thirty Initiatives that they have identified in this country, they hope to see 1,000 in five years' time. The Foundation is also working on a Local Initiatives support fund; and a technical advisory service in collaboration with I.T.D.G.

The Alternative Probation Project Day Centre

This project was established in 1976 by Jeananne Medd, a probation officer working in Farnham, Surrey. It was conceived as a creative alternative to the frustrating, conventional office and interview work that probation officers are obliged to perform when meeting young people needing emotional support and social education. Richard Cook has recently taken over the running of the project.

Essentially the idea is to work with a small group of young people in need of emotional support and social education, at a day centre which they attend for up to four months to learn to cope with themselves and organize themselves productively by undertaking a wide range of crafts, activities, and community service tasks – for example, a report on local drug-taking has been produced.

In 1978, forty-eight young people were referred to the project largely through the Probation Service. All were considered as high-risk re-offenders, most of whom had previous institutional experience. The project is strongly supported by the local probation office and receives independent funding from several charities.

The Land Trusts Association

The Association seeks to bring into wider public control and use large areas of private landholdings by the promotion of charitable land trusts. Charitable trusts will combine and safeguard the wider interests of the community and the individual owner's family. Through its Secretary, Joanna Abecassis, the Association will provide an initial advisory service for landowners and will promote the idea of putting land into trust for public benefit. It will stress in particular the need to create work opportunities in the countryside, and to maintain the vitality of rural communities. The group's greatest strength is its independence from the interests of central and local government, individual settlers and farming and landowning groups.

Greater London Association for Initiatives in Disablement

GLAID is a group chiefly of disabled people which aims to mobilize the disabled nationally to challenge the oppressive attitudes and institutions which at present characterize society's concern for the disabled. Organizer for the group is Micheline Mason.

The group aims to establish a movement towards self-help by setting up counselling services to help people with disabilities to overcome the negative stereotyping of society at large, and assisting them to set up home-based industries, small businesses, co-operatives, and craft workshops. A Fund for Initiatives will make small grants and loans to promising projects which need starter finance.

GLAID will continue to develop group work and other forms of support to help create more rational alternatives in work and attitudes.

The University of the Community

This is a collaborative initiative between the Open University and the Foundation for Alternatives. The focus of its activity will be the Group Project, concerned with local and/or national issues such as energy resources, rural development, local communications, delinquency and industrial organization. Students will attend full-time or on day release for two or three years, working on a suitable group project towards its completion as a co-operative effort.

The University initially will draw largely upon existing local resources and materials (school and college premises, the spare time of existing higher education staff, and local industry workshops and laboratories).

The Gulbenkian Foundation has offered a matching grant of half the sum required for initial adminstration. This has enabled a project manager, Rodney Stares, to be appointed to work on the delicate tasks of negotiating with the Open University and the Council for National Academic Awards about validation, identifying students, clarifying projects, seeking funds and forming support in the local community.

Associated Housing Advisory Services

AHAS has evolved since 1974 from the mutual interests and activities of its three member-directors in various national and international groups, working on local control and management of housing resources, including self-build housing.

In 1978 AHAS was incorporated as a non-profit making limited company whose prime purpose is: 'To provide support, assistance and advice to groups of persons within their local community who are seeking to improve their home and work environments in ways that will increase local and community control over local affairs; generate and retain income and wealth within the community; and conserve scarce or non-renewable resources.'

AHAS is supported by the projects it has generated and seeks to initiate only projects which have practical effect, rather than research for its own sake. AHAS projects include: studying a series of cases embodying policy principles involving users in planning, building and maintaining their own housing, which is part of a larger International Foundation for Development Alternatives (IFDA) project on housing.[11] This work has evolved into a second project: the designing of a Housing Tools Exchange Resources Index, a simple manual retrieval system mainly for Third World use. AHAS is now negotiating with five overseas groups planning to test the system for two years. Projects within the United Kingdom include: evaluating and monitoring a scheme of twelve self-build houses in Swindon, Wiltshire, and a similar sized scheme using simpler technology in Lewisham, south-east London. In addition, AHAS is making a study of the demand and constraints on self-build housing in the U.K. by interviewing local authorities, self-build groups and funding sources. The Directors involved in the AHAS project are Bertha Turner, Peter Stead and John Turner.

Institute for Community Care

As the public resources available for centralized health and social services diminish, pioneer schemes for community-based provision are being started by enlightened medical and social workers. In Barnsley, for example, Dr John Williamson and his partners have made their surgery into a base for mutual aid groups, and a source of education in responsible self-care. In Kent an imaginative Good Neighbours scheme is being run under the auspices of the County Social Services Department. In the village of Dinnington, near Sheffield, an action research project in community care is being supervised by the Department of Social Work of Sheffield University.

The Foundation's aim is to set up a national training centre, the Institute for Community Care, where the experience of these and other pioneers can be passed on to the wider body of practising professionals, either through special courses or by integration into their professional training.

Technology Choice

The last of these projects set up by the Foundation for Alternatives, in June 1979, is for work over a two-year period to develop the movement for technology choice in the U.K. The purpose is to raise the level of awareness among the public of the choices of technology available, and the likely consequences of the particular choices that are taken.

The method of working is to select key topics – agriculture, transport, health and housing – establish or reinforce working groups on these, clarify the nature of the choices which exist, and find ways of bringing these choices into the mainstream of public and private decision-making.

Environmental and Energy Groups

The Natural Energy Centre

The Natural Energy Centre, opened in 1977, is run by a group of engineers, technicians and administrators who aim to develop and promote a range of products which use natural energy as their primary source of power. N.E.C.'s role is totally practical: not only encouraging the development of new ways and methods of energy conservation, but also supplying natural energy and energy-sufficient equipment. The range of equipment is wide – solar collectors, heat pumps, wind-driven generators, wind pumps, methane digesters, wood and waste-burning stoves, electronic energy monitors, and waste heat recovery systems. The N.E.C. also provides a consultancy service which has been responsible for designing and supplying to schools and local authorities some of the largest solar panel installations in Britain.

The Natural Energy Association (N.E.A.), run from the N.E.C., is a forum for all those interested in natural energy, either theoretically or practically, and has a membership of over 3,000. A quarterly journal, *Natural Energy and Living*, carries features on energy and A.T.-related topics, a useful world 'round-up' column of innovations and applications of natural and non-nuclear energy systems, interviews, reviews and small advertisements. The Association also publishes a series of practical handbooks on natural energy; distributes the N.E.C. manual *How to Use Natural Energy*; operates the Natural Energy Book Service, and organizes symposia on natural energy.

Energy 2000 is a group campaigning for non-nuclear energy conservation policies, and the development of alternative energy sources, particularly the efficient and safe exploitation of U.K. coal reserves.

The Milton Keynes Development Corporation (p. 260) is not alone as a local authority taking an interest in alternative energy systems. In the past five years many county and district councils have expressed enthusiasm for new, and cheap, alternative energy systems, and a few have even gone as far as to install them in council offices and council-owned institutions. Some of the more recent local authority developments include: several council houses in Leeds fitted with solar panels to supplement the electric water heating system; eight Peacehaven (Sussex) council houses currently under construction incorporating a novel horizontal aerogenerator, the wind-wall and solar heaters; and a demonstration council house in Salford which uses three separate heat pumps and thick floor and wall insulation. A project partly organized through the local authority in Wandsworth has converted a row of fourteen houses to low-energy demand through extensive insulation and renewable energy and recycled waste energy techniques.[12] In Prestwick the Bury Borough Council are planning to build three council houses heated by solar energy and including extensive insulation at a cost of under £49,000, well within the standard allowance for local authority housing and with estimated heating costs one half those for conventional council houses.

The most ambitious scheme for low-energy demand council housing is on Humberside. Since 1977, Howard Liddell, an architect, and David Hodges, a physicist who was formerly at the Milton Keynes solar house, worked from the Hull School of Architecture to design thirty-two council houses which were built in 1980 on the edge of Hull. Both are founder members of the Alternative Technology Group at the Hull School of Architecture. The new designs incorporate high levels of insulation, and heating by means of individual storage tanks within each house, which are supplied by a single 100–130 kilowatt windmill. Supplementary heating will come from smokeless coal stoves (specially designed for the project by the National Coal Board). Overnight shutters and heat recovery from both ventilation air and waste water are being considered. As far as it is known, this is the only large-scale windpower project not designed to feed the national electricity grid. The emphasis in the project is on a high degree of self-sufficiency in both energy and food, and the layout of the terraced houses caters for fairly large gardens suitable for growing vegetables.[13]

This unique project has attracted widespread backing and interest, notably from the local authority, which has approved a 10 per cent addition to the standard costs of the council houses to cover some of the modifications. Finance over this basic grant is currently being sought from the Energy Technology Support Unit based at Harwell (headquarters of the U.K. Atomic Energy Authority) for the technical aspects of the work being done. The Nuffield Foundation has allocated £8,000 to support a study into the social aspects of the project – the reactions and adaptability of the tenants to their novel 'alternative' housing.

Although Hull Council have a firm political and financial commitment to this project, it could suffer considerable delays due to the insistence on the part of the Department of the Environment that unconditional legal and financial commitments are made by the N.C.B., Hull Council, ETSU and others involved in the project, before they approve the allocation of government funds to finance the houses. These bodies are themselves awaiting Department of the Environment approval before they enter into such agreements. The DoE feels that the project tries to cover too many aspects, the management of which would prove difficult.

Howard Liddell and the A.T.G. have also been working with a local group on a rural self-sufficiency study project in Aberfeldy, in the Upper Tay Valley, Perthshire. This followed the widespread notice which a preliminary self-sufficiency strategy for the valley, drawn up in 1974 by local residents, attracted in the press – notably *Farmers Weekly*, the *Ecologist*, and certain regional and national newspapers. A group of local people, the Provost, county councillors, the local doctor, tradesmen and farmers expressed a common concern about the U.K.'s deteriorating economic position in the wake of the 1974 OPEC oil price rises, the collapse of the fatstock market at the end of 1974 and future world food shortages. They were very worried too about the rising costs of transporting their agricultural produce to distant, centralized markets, only to have it reimported as processed foods and materials for local consumption. They felt that the economic dominance exerted by these distant industrial centres grossly distorted rational and sustainable land and other resource use, and also the range of employment opportunities available in the valley. Their response was to suggest that the valley should become as far as possible self-sufficient in the commodities it used.

The A.T.G. agreed to provide comprehensive feasibility studies of self-sufficiency in food and energy in the valley. Their report made clear that with a higher level of formal and informal local economic activity in agriculture and agricultural processing, the population in Aberfeldy Burgh (7,220) could become largely self-supporting, and this would involve no deterioration in the current standards of living.[14] These studies are being developed further by the locally based Achloa Institute, where Howard Liddell now lives. Detailed work is under way on different aspects of community life – health, education, recreation and so on. This imaginative programme aims, first, to create a greater diversity of agriculture, developing local technology and local markets; and secondly, to use tourism creatively, by 'presenting' the area as a working, living entity exhibited on several sites – showing farming, health care, food processing, forestry and industry – and the potential for a self-sufficient community.

Friends of the Earth

Friends of the Earth Ltd (F.O.E.) is the best known and certainly the most widely respected apolitical pressure group advocating rational resource use policies and less rapacious lifestyles. Formed in the U.K. in 1971, it has over 10,000 members who are kept in touch through regular newsletters with the progress in F.O.E.'s national campaigns to direct public attention to the environmental implications of industrial and social planning, and to stimulate public policy changes in land use, transport, energy, wildlife, natural resources and bicycles. Earth Resources Research Ltd, the research branch of F.O.E., has published a series of first-class reports – a recent one dealt with the likely impact of microprocessors on employment in Britain – and a number of expertly written books related to F.O.E. campaigns including Walter Patterson, *The Fissile Society* (1977), Catherine Thomas, *The Paper Chain* (1977), and Colin Hines and Graham Searle, *Automatic Unemployment* (1979).

There are about 260 autonomous local F.O.E. groups throughout the U.K., from Cornwall to the Shetlands, which are working on local applications of F.O.E. policies and which take a very active part in local campaigns – local government planning, waste recycling and anti-nuclear power lobbying – as well as contributing valuable effort to the national F.O.E. campaigns.[15] About thirty of these groups run full-time local F.O.E. offices supported through sales of F.O.E. publications and publicity materials, and through other commercial enterprises such as paper recycling, insulation projects and wholefood shops. F.O.E. Birmingham Ltd operates a most impressive work centre in a very large refurbished warehouse in the centre of the city. A dozen people are employed full-time in the co-operative running of the F.O.E. office; organizing paper recycling; selling F.O.E. recycled paper and other locally produced F.O.E. products; and in managing the Environmental Information Project and a Housing Insulation Project for the elderly and disabled. Part of the warehouse is given over to a wholefood wholesale outlet and the production of *Wholefood News*, the magazine of the Wholefood School of Nutrition.

Insulation schemes are run by at least ten other local F.O.E. groups – for example, in Portsmouth, Canterbury and Southampton. The Southampton group also runs a paper recycling project (which employs ten people), as do the groups in Bournemouth and Kensington and Chelsea. Wholefood shops started up by F.O.E. groups are to be found in Durham, Bedford and Southampton; and in Bristol F.O.E. has lately begun an Environmental Education Project under an M.S.C. scheme.

Other organizations which mobilize people and information on environmental, resource and energy issues include the Conservation Society, with over fifty very active local branches working for sustainable and equitable natural resource use policies; the Society for Environmental Improvement, which, amongst other activities, sponsors the Centre for Alternative Technology; and the Political Ecology Research Group. There are also a number of environmental information and education services dotted around the country – for instance, the Environmental Groups Information Service (EGIS) in Newcastle, and the Environmental Information Centre in Nottingham, started in 1977 by the local branches of F.O.E. and the Conservation Society. A comprehensive annual Directory of Environmental Groups is published by the Civic Trust.

The Green Alliance is a pressure group formed in 1979 by Lord Beaumont to canvass political support for environmental issues within the House of Commons and the House of Lords, and impress upon politicians of all parties the need to devote greater consideration to matters concerning the natural environment in forming public policy. Their weekly 'Parliamentary Newsletter' contains full information on relevant ministerial statements, membership of committees, publications and inside profiles on ministers and their degree of interest in all matters 'green'. An active group within Parliament itself is the Parliamentary Liaison Group for Alternative Energy Strategies (PARLIGAES), which has recently appointed a Parliamentary Liaison Officer with the aim of keeping M.P.s as well informed about renewable sources of enegy as they now are about conventional energy policies. It is also intended to keep them informed about the social, ecological and other implications of the conventional policies and the alternatives to them.

Numbered amongst the supporters of the three main political parties are members of very lively, self-supporting and unofficial pressure groups which are busily urging their Parliamentary representatives and other party members to endorse sustainable, ecology-oriented resource use policies. The Conservative Ecology Group, formed in 1977, produces a series of short papers on ecology and resource studies which are distributed to Conservative M.P.s, and contributes to discussions on Conservative policy in an attempt to make conservation and environmental issues acceptable to the party.

Also formed in 1977, the Liberal Ecology Group participates in a well organized campaign supported, in spirit at least, by the majority of Liberal Party members to press Liberal M.P.s to adopt environmentally sound policies based on ecological principles. Although A.T. does not form part of the official programme of any of the political parties, most Liberal Party politicians are sympathetic advocates.

By far the most vigorous of the political pressure groups is the Socialist Environment and Resources Association (SERA), set up in 1973 to involve the Labour movement in the debates and decision-making processes concerned with global shortages of raw materials and food, and the environmental and human degradation caused by large-scale technology. SERA is supported by at least fifteen Labour M.P.s, and backed up by a well informed monthly newsletter and regular topical pamphlets. The Association has numerous local branches and specialist working groups on energy, microprocessors, employment and transport. With many local contacts in the trades union and co-operative movements and the Labour Party, SERA has been energetically promoting the workers' plans at Lucas Aerospace, working with CAITS (see pp. 99–100), and in the power engineering industries – especially proposals for combined heat and power systems – and providing invaluable constructive help to local Co-operative Development Agencies and co-operative projects. SERA members played a prominent part in drafting the Labour National Executive Committee's 'Statement on the Environment', accepted by the 1978 Labour Party Conference.

Concern about a sustainable future gave rise to the Ecology Party, founded in 1973 following the publication of *Blueprint for Survival*.[16] In the general election of May 1979 the party fielded fifty-three candidates and polled nearly 42,000 votes, making it the most successful minority party in the country. Despite winning no seats in that election the Ecology Party has had notable successes in county, district and parish elections, arguing in favour of policies for an increasingly self-sufficient, sustainable national economy, based on the 'economics of permanence', particularly in food and energy. Regional policies focus on programmes for greater self-reliance in housing, transport and employment based on low environmental impact, and decentralized, small-scale industries. The party is firmly committed to strict environmental protection measures; the devolution of decision to regions, and neighbourhoods; and radical policies on pollution and health.

Alternative Health Care

The Institute for Community Care (see p. 265) is planned as a professional training centre which builds on the experience of the growing number of self-help schemes and organizations devoted to setting up alternatives to the conventional doctrine of health care. The Scientific and Medical Network (S.M.N.) is an informal group consisting mainly of qualified scientists and doctors, many of whom are active in education. The S.M.N. seeks to redress the purely scientific materialist outlook of the orthodox scientific professions by informally introducing and incorporating into research and training programmes a broader awareness of the intuitive and spiritual insights that stem from a more holistic approach to life.

The Healing Research Trust, a charity, aims to inform and advise the public on the benefits of alternative therapies such as acupuncture, herbalism, homeopathy, etc., and to encourage practitioners of all therapies to work together. The Trust campaigns for the public's right to have the treatment of their choice, and for the national establishment of Natural Health Centres where qualified practitioners from several therapies would be available to give advice and training in self-help.

Within orthodox medicine the Unit for the Study of Health Policy based at a major London teaching hospital, Guy's, is drawing attention to the severe health costs of indiscriminate economic growth – a major contemporary hazard to public health. The Unit sees many advantages for health in an ecologically sound pattern of economic development and is seeking to promote an understanding of these benefits.

Town Planning

Quite apart from the growing number of people who are involved in self-build housing projects, imaginative proposals have been put forward as the result of the activities of groups interested in unconventional housing and planning schemes.

The Town and Country Planning Association (T.C.P.A.) is a voluntary organization founded in 1899 to promote the ideas on decentralized urban living set out by Ebenezer Howard in his book *Garden Cities of Tomorrow*. In 1903 Howard and his friends started building Letchworth in Hertfordshire as the first garden city. (It is the only town in Britain where, by law, the surplus income of the Garden City Corporation is spent for the benefit of the citizens.) After the First World War the second garden city was founded at Welwyn, and following the Second World War the New Towns Act 1946, made the development of new towns official policy. The T.C.P.A continues to advocate dispersal and its other activities include a planning-aid service and an education service conducted through its *Bulletin of Environmental Education* (*BEE*). Currently the T.C.P.A is involved in negotiations with the Milton Keynes Development Corporation over the plans and details of the 'Third Garden City' project (see the Open University Alternative Technology Group, p. 260).[17]

The Building and Social Housing Foundation (B.S.H.F.) is an educational and research organization dedicated to experimenting in and disseminating information on new developments in the construction and management of residential housing. The B.S.H.F. works in close co-operation with the International Co-operative Housing Development Association and the United Nations Human Settlement Foundation.

The most spectacular scheme with which the B.S.H.F. has become associated is Konrad Smigielski's plan for a co-operative, self-sufficient community set in sixty-five acres of the grounds of Stanford Hall, a college for the co-operative movement, near Loughborough in the East Midlands. A population of 1,300 is being planned for in two adjoining residential districts – one 'semi-rural', the other 'urban' – both using energy from the sun and the wind. The proposal is

> to create a new, ecologically balanced semi-rural community with a definite role to play in the economic and social life of the

countryside. The village will be based on energy-conserving technologies . . . Recycling of waste and the use of salvaged building materials will be part of our techniques, based on the principle that nothing is destroyed or wasted but only transformed, re-used or returned to the land.[18]

The plan also allows people to design and build their own houses.

Other projects undertaken by the B.S.H.F. include research and design for restoring a run-down farm in Charnwood Forest, Leicestershire, for use as a research – self-sufficiency and specialist agricultural projects – and educational centre; assessing the progress and development of a self-help housing project in Derby; and, in collaboration with the National Foundation of Housing Associations, a programme to examine and devise alternative sources of housing finance.

Society, Religion and Technology

A remarkable project which has been running for nearly a decade, and which has covered many of the activities currently being examined by alternatives groups, was started in Edinburgh by the Church of Scotland. The Society, Religion and Technology Project, founded in 1970, originated from feelings of disquiet within the Church of Scotland, that it would become unable to respond to human problems within a rapidly developing technological world if it remained ignorant of the reasons underlying how and why technological decisions were made. The Church was also aware that certain types of technology – such as nuclear power – posed very profound ethical and theological questions. The project was established to:

define the problems and investigate the opportunities raised by the development of new technologies;

co-ordinate the study of theoretical and practical issues emerging from these assessments;

identify the specific challenges to Christian theology and to involve technologists and theologians in constructive dialogue;

develop methods of regularly reporting this work at as many levels as possible throughout the Church.

Studies published by the project on alternative energy sources, ecology and ethics and direct practical work with a solar energy housing project in Iona have been at a consistently high level.[19] The project established a number of respected study groups and produced audio-visual materials dealing with natural resource and food shortages and their implications for the Christian faith.

Following the departure of Colin Pritchard, Director of the project until 1978, Ian McDonald has taken over the running of many existing programmes – the Iona Solar Energy Project and assessment of renewable energy sources in the Highlands, jointly undertaken with the H.I.D.B. – and introduced new schemes.

A church group from Heidelberg, Germany, are very interested in helping set up an A.T. centre in Scotland and are gathering information on A.T. groups in Germany and Holland prior to visiting U.K. groups for background information on how a Scottish A.T. centre might develop.

Ian McDonald is starting up multidisciplinary studies in unemployment and technology, medical resources for the elderly and energy production in order to give a broad, overall approach to these matters that fits in with the Church of Scotland's design for the project. A wide range of contacts within local F.O.E. groups, the S.D.A., the World Development Movement, trades unions, the W.E.A., community groups, and the Church itself also facilitates a comprehensive educational programme within schools and the community.

The Literature of Alternatives

Apart from the newsletters of individual groups there are a number of flourishing information exchanges, newsletters and network groups serving the A.T. and alternatives movement. Recently revived after three years in abeyance is the Network for Alternative Technology and Technology Assessment (NATTA: c/o Faculty of Technology, Open University, Milton Keynes), which held a conference entitled Alternative Technology: Institutional Co-option or Local Control? at Milton Keynes in August 1979.

NATTA provides a forum for debate and discussion on topics of
A.T. in Britain, including information on A.T. training courses
and careers.

The 1979 COMTEK (Community Technology) Festival was held to
coincide with the NATTA conference, and, like the COMTEK Festivals
in 1974, 1975 and 1976, there were exhibitions of A.T. hardware
and demonstrations of techniques with an emphasis on D.I.Y. and
community-scale technology, plus the usual stalls, films, sideshows
and theatre groups. Following the success of the festival a co-
operative registered under the name of COMTEK was set up. Over
the next few years it will build up a product line of A.T. manufactures and services – bicycles and tricycles, woodburning stoves
and a domestic and industrial insulation service. The co-operative
also intends to draw upon A.T. research within the Open University as the basis for new products and services.

Although the media are generally sympathetic to alternatives,
the only newspaper to feature the subject regularly is the
Guardian, in which Harford Thomas's weekly column on alternatives is truly invaluable.

Turning to the regular network publications providing round-
ups of A.T. group activities, a very well informed newsletter,
distributed monthly to over 1,400 people mainly in the U.K., is
produced by Roland Chaplain and his colleagues at the Futures
Study Centre (F.S.C.: 15 Kelso Road, Leeds LS2 9PR). For the
past two years the F.S.C. has run a course on Alternative Lifestyles and Appropriate Technologies covering varied topics like
co-operatives, renewable energy sources and the Lucas workers'
plan. Future courses are likely to focus on the regional alternatives
movement, Alternatives for Leeds.

One of the most effective and most widely known A.T. newsletters is James Robertson's *Turning Point* (9 New Road, Ironbridge, Telford, Salop TF8 7AU), which brings together some
1,500 people, mainly in Britain, North America and Europe who
share a common interest in the alternatives movement and its
world-wide development. *Changing Direction* is an international
project broadly designed to facilitate the exchange and dissemination of useful information and ideas about the social, economic
and political guidelines for an alternative future for industrialized
society. The nature of the project is outlined in the final chapter of
James Robertson's recent book, *The Sane Alternative*.[20]

To date, the project has engaged in a wide range of activities (including lectures, articles and seminars for senior business people and public officials) and has begun to build up a wide range of information and contacts relating to the strategy for a 'sane, humane and ecological' future.

For the past four years John Davis has been working on a project looking at many aspects of 'A.T. for Britain' and has been responsible for what has become known as the A.T./U.K. Unit of I.T.D.G. His quarterly newsletters, *A.T./U.K. Exchange* (10 Grenfell Road, Beaconsfield, Buckinghamshire), contain a very useful mix of comment, analysis and description of the alternatives movement.[21]

Mainly serving the thriving commune sector is the Communes Network (Lauriston Hall, Castle Douglas, Kirkcudbrightshire), which provides much valuable news, and regular bulletins of events, practical courses and skill exchanges throughout the country.

There are about eighty alternative, community newspapers published by local groups that generally pick up local news items which are neglected by the conventional press. They provide valuable information about local alternatives groups, women's groups and co-operatives. People's News Service (Oxford House, Derbyshire Street, London E2 6HG), provides a fortnightly national and international summary of articles in community newspapers from many part of the world, and draws together stories with common themes.

In an effort to improve library collections of A.T. materials, the Alternative Technology Information Group (A.T.I.G.: 148 Lovelace Road, Surbiton, Surrey) was set up in 1976. This group of information workers and librarians aims to help public and other libraries build up their collections of A.T. literature and broaden public access to it. A quarterly newsletter to members gives information on A.T.I.G. activities such as classification, a thesaurus, bibliographies and so on, and details of new books and organizations in the field.

In 1980, A.T.I.G. started to produce a categorized directory containing brief annotations of A.T. literature, periodicals and organizations, plus a list of government bodies to approach for A.T. funding. The aim is to produce a U.K. equivalent of the American *Rain Book*.[22] A.T.I.G. has also participated in the setting up of the *A.T. Index*, a quarterly index of books and periodicals covering the literature on appropriate technology and many related fields.

The main regular journals devoted to alternatives (as distinct from newsletters which are generally restricted to members of a particular organization) are *Undercurrents*[23] – undoubtedly the most informative; *Vole*; *Whole Earth*; *Practical Self-Sufficiency* – which is *the* practical man's guide to every aspect of self-sufficiency; *Ecologist*; *Resurgence*; and (chiefly for developing countries) I.T.D.G.'s *Appropriate Technology*.

There are also several directories to co-operative and A.T. groups:

In the Making (*I.T.M.*: 44 Albion Road, Sutton, Surrey) is a non-profit making, national directory of co-operative projects ranging from small, fairly conventional firms to small 'communal' collectives. The projects covered have two things in common; they are productive, i.e., provide long-term ways of making a living, and co-operative in the sense that they practise shared decision-making.

I.T.M. concentrates on projects just getting off the ground but also covers firmly established and developing ventures. The directory also contains short articles on the theory and practice of collective working; descriptions of the aims and status of individual co-operative projects; information on sympathetic organizations, publications and groups; a co-op 'small ads' section; a section of book reviews and events, plus a detailed index. The directory appears annually and the latest edition (1980), *I.T.M.* 7, contains details of over 100 co-operatives and sixty-three support and advice groups in the U.K.

Although *I.T.M.* does not claim to be a comprehensive directory of co-operatives and community projects (it relies upon individual projects making themselves known to *I.T.M.*), it is specific and well put together. It provides detailed information on the development of small co-operatives and the alternatives movement in the U.K.

A lively new quarterly on co-operatives was started in 1980, *Co-operative Times*, edited by John Parkinson, who runs a research and consultancy group in Solihull, Birmingham.

Ways and Means is a one-off national directory of projects, groups, organizations and publications that are in some way connected with the radical alternatives movement in the U.K. The directory, published by the National Union of Students, is much broader in scope than *In the Making* and lists organizations active in the field of minority rights, women's rights, the media, education, housing, co-ops, community action and volunteer work.

The *Alternative Technology Directory* (11 George Street, Brighton BN2 1RH), provides a wide ranging index of organizations concerned with A.T., resources, energy, health and safety, rural environment and the Third World. The index covers articles published since 1975 in the main 'alternative' journals on the topics listed above: a useful guide for searching out bygone articles and little known texts.

NOTES

1 The Formation of the I.T.D.G.

1 'Reflections on the Problem of Bringing Industry to Rural Areas' (Planning Commission, New Delhi, 1962). Copies of this report are available from I.T.D.G. on request.
2 Published in *Roots of Economic Growth* (Gandhian Institute of Studies, Varanasi, India, 1962).
3 Blond and Briggs, London, 1973. Harper & Row, New York, 1973.
4 *Roots of Economic Growth*, op. cit., p. 9 (first published in the *Observer*, Weekend Review, London, August 21st, 1960).
5 Ibid., p. 49.
6 Appropriate Technologies for Indian Industries, held at the Small Industry Extension Training Institute, Hyderabad, January 1964. A report of the conference was published by S.I.E.T. Institute, 1964. Those present included T. N. Singh; Annasaheb Sahasrabudhe; and D. K. Malhotra, Planning Commission; Professor D. R. Gadgil, Gokhale Institute, Poona; R. N. Jai and J. E. Stepanek, S.I.E.T. Institute.
7 Included in R. Robinson (Ed.), *Developing the Third World: the Experience of the 1960s* (C.U.P., Cambridge, 1971).
8 Ibid., p. 4.
9 Extract from an interview with E. F. Schumacher, Radio Station K.P.F.A., California, September 1974.
10 *Observer*, Weekend Review, August 29th, 1965.
11 E. F. Schumacher, *Good Work* (Cape, London, 1979. Harper & Row, New York, 1979).

2 The Work of the Group Today

1 James Keddie and William Cleghorn, 'Least Cost Brick-making', *Appropriate Technology*, Vol. 5 No. 3, November 1978. The details of this research are given in a book by these two authors, *Brick Manufacture in Developing Countries*, one of the series *Choice of Techniques in Developing Countries* (Scottish Academic Press, Edinburgh, 1980).

2 *Mobilizing Technology for World Development*, Report of the Jamaica Symposium, International Institute for Environment and Development, March 1979.

3 I.T. Publications, revised edition, 1974.

4 I.T. Publications, revised edition, 1974.

5 I.T. Publications, 1976.

6 Rodale Press, Emmaus, Pennsylvania, 1980.

7 See I.T. Publications list, January 1979.

8 For instance, at the International Institute for Tropical Agriculture, Ibadan, Nigeria, on minimum tillage systems; and the East-West Centre's INPUTS (Increasing Productivity Under Tight Supplies) project, due for completion in 1979. For examples in a rich country, see pp. 144–6.

9 See Lester R. Brown, *The Global Economic Prospect: New Sources of Economic Stress*, Worldwatch Paper 20, Worldwatch Institute, Washington, D.C., May 1978.

10 I.T. Publications, 1975.

11 Lester R. Brown, op. cit.

12 The extent to which doctors are ensnared and their values distorted by increased specialization and advanced techniques is nowhere better illustrated than by the birth of the second test-tube baby in the world in Calcutta, a city where children are dying on the streets every hour. See *The Times*, October 6th, 1978.

13 K. Elliott, 'Using Medical Auxiliaries' in O. Gish (Ed.), *Health Manpower and the Medical Auxiliaries* (I.T. Publications, 1971).

14 Ibid.

15 K. Elliott (Ed.), *The Training of Auxiliaries in Health Care* (I.T. Publications, 1975).

16 K. Elliott (Ed.), *Auxiliaries in Primary Health Care* (I.T. Publications, 1979).

17 M. Skeet and K. Elliott (Eds), *Health Auxiliaries and the Health Team* (Croom Helm, London, 1979).

18 This scheme has now been included in the W.H.O./UNICEF joint study on alternative approaches to meeting basic health needs in developing countries.

19 Trevor Bottomley, *Business Arithmetic for Co-operatives and Other Small Businesses*; B. A. Youngjohns, *Co-operative Organization*; Peter Yeo, *An Initial Course in Tropical Agriculture for the Staff of Co-operatives*, 1976. All published by I.T. Publications.

20 Marilyn Carr, 'Appropriate Technology for Women', *Appropriate Technology*, Vol. 5 No. 1, I.T. Publications, 1978.

21 Marilyn Carr, 'Appropriate Technology: Its Importance for African Women', paper read at Karen College, Nairobi, October 1977.

22 See Carl Riskin, 'Intermediate Technology in China's Rural Industries', *World Development*, Vol 6, No. 11/12, 1978 (Pergamon Press, Oxford).

23 Personal communication from Anthony Ellman, Commonwealth Secretariat, May 1978.

24 See N. Jéquier, *Appropriate Technology Directory* (O.E.C.D. Development Centre, Paris, 1979), which lists over 200 A.T. groups in developing countries.

3 Technology: the Critical Choice

1 Milton R. Freedman (Ed.), *Intermediate Adaptation in Newfoundland and the Arctic* (Institute of Social and Economic Research, Memorial University of Newfoundland, 1969, republished 1974).

2 Unwin University Books, London, 1963. Beacon Press, Boston, 1969.

3 See especially E. F. Schumacher, *Good Work* (Cape, London, 1979. Harper & Row, New York, 1979); Hazel Henderson, *Creating Alternative Futures* (Berkeley Publishing Corporation, New York, 1978); and James Robertson, *The Sane Alternative* (River Basin Publishing Co., Minneapolis, 1979).

4 See *Financial Times*, December 15th, 1978.

5 A study published in 1979 shows that two-thirds of all new jobs in the U.S.A. come from companies employing fewer

than twenty people. See David L. Birch, *The Job Generation Process* (Massachusetts Institute of Technology, 1979).
6 E. F. Schumacher, from an unpublished paper dated 1976.
7 Address to the Conference on Appropriate Technology for the U.K., Newcastle upon Tyne, March 1976.
8 Hazel Henderson, op. cit., p. 386.

4 The Alternatives Movement in Britain

1 See Sir Kenneth Blaxter, *Energy Use in Farming and its Cost* (Rowett Research Institute, Aberdeen, 1978).
2 Gerald Leach, *Energy and Food Production* (I.P.C. Science and Technology Press, London, 1976).
3 E. F. Schumacher, 'The Next Thirty Years', *Soil Association Quarterly Review*, December 4th, 1977.
4 The Royal Commission on Environmental Pollution is far from reassuring about the environmental effects of conventional chemical farming. See its Seventh Report, *Agriculture and Pollution* (Cmd. 7644, H.M.S.O., London, 1979).
5 Biological soil husbandry is defined as those techniques of soil care and cropping which result in the efficient utilization of the sun's energy and of dependent biological processes for sustainable food production, using local resources alone.
6 Written by R. D. Hodges for the I.I.B.H., 1978, and distributed by the Henry Doubleday Research Association, Bocking, Braintree, Essex.
7 These include the Soil Association, Country College, the Henry Doubleday Research Association, the Centre for Appropriate Technology, the International Association of Organic Gardeners, and W.W.O.O.F. (Working Weekends on Organic Farms). COMET is at Well Hall Country College, Alford, Lincs.
8 Besides the organizations already referred to, many relatively new groups are working on different aspects of organic husbandry. These include the Centre for Appropriate Technology, New Mills, the Bridge Trust, the New Villages Association, the Commonwork Land Trust and the Rural Resettlement Group. The organic smallholder is very well

served with information by *Practical Self-Sufficiency*, a bi-monthly journal published from a smallholding in Saffron Walden, Essex.

9 See also an excellent booklet, *Meanwhile Gardens*, by Jamie McCullough, which gives a practical description of setting up a garden and recreation centre in Paddington, London (Calouste Gulbenkian Foundation, London, 1978).

10 *The Future of Employment in Engineering and Manufacturing*, CAITS, North East London Polytechnic, 1979.

11 *Training for Skills*, Manpower Services Commission, 1977.

12 An excellent account of this is given by David Elliott, *The Lucas Aerospace Workers' Campaign*, Young Fabian Pamphlet 46, Fabian Society, 1977. The remainder of this section is largely based on his work.

13 Lucas Aerospace Ltd is a division of the British-based multinational firm, Lucas Industries Ltd, and produces a wide range of electrical and mechanical systems and precision components for the aerospace, automobile and engineering industries. Generally the company deals in small batch, advanced technology production of specialist accessories primarily for the aerospace industry, rather than mass production. Roughly half of its work is in the field of armaments.

14 Although the company has never recognized the Combine as a negotiating or consultative trades union body, a few meetings have been held between company representatives and the Combine on certain trades union issues.

15 It is important to make the distinction between the trades union activities of shop stewards and the official policy of the trades unions as espoused by their national leaders. The actions of shop stewards, who are based firmly at the point of production (or service), are frequently independent of and generally more advanced than the decisions and policies given public utterance by the trades union leaders. The Combine is recognized by the Lucas Aerospace shop stewards but not by the thirteen trades unions which have a place on the Combine Committee. The company is obliged to follow the official trades union line and treat the Combine as 'unofficial'.

16 *New Scientist*, March 21st, 1974.

17 The anti-human characteristics of new processes are not confined to the machinery used by those workers actively in-

volved on the shop floor. They are becoming a dominant
feature of the hardware used by the professional and clerical
staff in all manner of firms. See M. J. E. Cooley, 'The New
Proletarians', *New Scientist*, August 24th, 1978; M. J. E.
Cooley, 'Taylor in the Office', in R. N. Ottaway (Ed.),
Humanizing the Workplace (Croom Helm, London, 1977).

18 M. J. E. Cooley, 'Design, Technology and Production for
Social Needs', in Ken Coates (Ed.), *The Right to Useful Work*
(Institute for Workers' Control, Spokesman Books, Notting-
ham, 1978).

19 M. J. E. Cooley, 'Design, Technology and Production for
Human Needs', in Coates, op. cit.

20 *Lucas: An Alternative Plan*, Institute for Workers' Control
Pamphlet No. 55, 1977.

21 D. Elliott et al., *Trade Unions, Technology and the Environment* (O.U.P., Oxford, 1978).

22 Before it was complete, several of the ideas were used to fend
off redundancy threats in June and October 1975 at the
Marston Green (Birmingham) and Hemel Hempstead
(London) plants. Mini-corporate plans were quickly drawn up
by the shop stewards and in both cases redundancies were
averted. They won, according to *The Economist* (October
4th, 1975) 'by showing management how to manage'.

23 *The Corporate Plan*, L.A.C.S.S.C., 1976.

24 A full summary of the proposals can be found in *Lucas: An
Alternative Plan*, op. cit.

25 See *New Scientist*, July 3rd, 1975.

26 The Combine did not, however, release specifications of all
150 products. Instead they presented a summary of the proposals and the section on alternative energy technologies gave
the details of only twelve products. They were worried that if
the entire Plan were released the company might pick out the
most profitable items and reject the non-commercial, socially
useful products which make up nearly half the product range.
To ensure that the products selected for presentation were not
introduced piecemeal in only one or two sites, they chose
those which could only be manufactured by the co-ordinated
effort of several factories.

27 *Lucas: An Alternative Plan*, op. cit.

28 These tended to cut across traditional skill and job demar-

cation, wage differentials and so on. Such radical changes within the accepted structure of the manager/employee cash-nexus and in the relationships between individual trades unions needed to be made with care and not forced through in great haste.

29 The management argued that the existing product range was profitable, that it would remain so and that all their products had undeniable social utility. Their only concession was a proposal for talks at individual sites to review alternative products, but not necessarily those in the Plan. The company went on to make it plain that they were unwilling, in these or any other consultations, to work with the Combine Committee. It was not regarded as part of the recognized consultative machinery.

30 Geoffrey Foster, 'Lucas Aerospace: The Truth', *Management Today*, January 1979.

31 At several sites shop stewards began to press for parts of the Plan to be implemented. Mass meetings and discussions between workers and management were held to discuss the proposals.

At the Burnley, Lancashire, plant a meeting of over 200 people was held to review the Plan in the light of the company's statement. Several local Lucas managers, site employees, local trades councils, workers from other local factories and members of local community groups attended the meeting. It proved a success, with both managers and employees involved in discussing details and refinements of the Plan.

The management agreed to introduce a development programme for heat pumps and to study other alternative products. In 1977, in conjunction with the Energy Research Group of the Open University, work began on two prototype, small-scale, natural gas powered heat pumps. (These are currently being field tested in council houses in Milton Keynes.)

32 See L.A.C.S.S.C., *Democracy versus the Circumlocution Office*, Institute for Workers' Control, Pamphlet No. 65, 1979. Their attempts to win the backing of trades unions have made little headway in the face of the antipathy of these organizations to the 'unofficial' Combine.

The Labour Party expressed support for the proposals and over forty Labour M.P.s have been active proponents of the Plan since 1976. However, the DoI, under the Labour administration, though they publicly welcomed the Plan, in practice ignored it. Lucas Aerospace management has been consistently hostile.

33 A conference, 'Alternatives to Unemployment', held jointly by CAITS and N.E.L.P. in November 1978, attracted 500 participants from trades unions, universities, local authorities and overseas. Twenty background papers from the conference were published by CAITS. Two recent (1979) publications are *The Future of Employment in Engineering and Manufacturing* and *Energy Options and Employment*.

34 While these were being drawn up, in September 1978, A.T.V. put out two hour-long programmes on the Lucas campaign. Both were notable campaign victories for the Combine. But see Mike George, 'Behind the Scenes: How A.T.V. handled the Lucas Aerospace Affair', *Workers' Control*, Bulletin No. 3, I.W.C., 1979.

35 Available from CAITS.

36 The working party is still considering its recommendations (1980).

37 However, the Combine shop stewards are far from happy with the composition of the working party. The C.S.E.U. has never been a committed supporter of the Plan or the Combine. Some observers argue that the relationship between the company and the C.S.E.U. is far too cosy: see 'Combine News', *Workers' Control*, Bulletin No. 3, I.W.C., 1979.

38 At first it was argued that the Lucas campaign was exceptional. Only a few firms in the U.K. combine the same elements of production as Lucas Aerospace – a highly skilled workforce and a highly adaptable technology working on the one-off or small batch manufacture of very specialized products. The variety of industries which have taken up Lucas-style campaigns indicates that such reservations were misplaced. Although none of the campaigns has so far been as successful as that at Lucas, they have yet to develop the broad experience and solid shop-floor support that is the hallmark of the Lucas Combine. In a number of cases they have been started up too late and redundancies have foreclosed them.

Pre-emptive campaigns were therefore not possible and in some cases had not even been considered.

39 *Workers' Control,* Bulletin No. 32, I.W.C., May/June 1976.

40 C.S.E. Books, 1979 (55 Mount Pleasant, London, WC1X 0AE).

41 Production of the twenty-year lifetime car has also ·been suggested.

42 Copies may be obtained from Vickers North East Working Group, c/o Benwell C.D.P., 87 Adelaide Terrace, Newcastle upon Tyne 4. See also Huw Beynon and Hilary Wainwright, *The Workers' Report on Vickers* (Pluto Press, London, 1979).

43 The Scotswood site in Newcastle was closed by the management in May 1979. Eight hundred workers lost their jobs. See Mary Kaldor, 'How arms sales trigger decline', *Guardian,* May 7th, 1979.

44 *Dunlop: Jobs from Merseyside – A Trade Union Report* (CAITS, N.E.L.P., London, 1979).

45 *Military Spending, Defence Cuts and Alternative Employment* (T.G.W.U., 1977).

46 American Federation of State, County and Municipal Employees.

47 *Report of the Committee of Inquiry on Industrial Democracy* (H.M.S.O., London, 1977).

48 The background to worker co-operation in the U.K. and Europe and the arguments for an extension of the principle in this country have been eloquently put by Robert Oakeshott in *The Case for Workers' Co-ops* (Routledge and Kegan Paul, London, 1978).

49 A credit union is a co-operative which provides its members with an efficient, inexpensive savings – and loan – service. See Jack Dublin, *Credit Unions* (Wayne State University Press, Detroit, 1971).

50 However, it would be unfair to give the impression that the C.P.F. model rules are inflexible and without certain advantages. As Paul Derrick, Secretary of the C.P.F., has pointed out, the model rules can be amended so as to restrict shareholdings and the distribution of assets and raise or lower the financial limits on shareholding. Moreover, their detail and structure, which can be adjusted according to requirements,

makes them very suitable for enterprises involving many in-
dividual co-operators. (See Paul Derrick, 'The Co-operative
Productive Federation', *In the Making* 7, 44 Albion Road,
Sutton, Surrey.)

51   For an excellent account of the background and struggles to
     set up each of these co-operatives see Ken Coates (Ed.), *The
     New Worker Co-operatives* (I.W.C., Spokesman Books,
     Nottingham, 1976).

52   Martin Leighton, 'The Workers' Triumph', *Sunday Times
     Magazine*, June 4th, 1978.

53   *Guardian*, December 12th, 1978.

54   *Manpower Services Commission Review and Plan 1977*
     (M.S.C., London, 1977).

55   Tony Eccles, 'Kirkby Manufacturing and Engineering', in
     Ken Coates (Ed.), op. cit.

56   Tom Lester, 'The Crumbling Co-ops', in *Management Today*,
     February, 1979.

57   Common ownership has been defined as the 'ownership and
     control of an enterprise by (at least the majority of) those
     working on it. The supply of money to a common ownership
     enterprise gives no right to any participation in ownership,
     control or profits.' Alastair Campbell, *Worker Ownership*
     (ICOM, London, 1973).

58   See Susanna Hoe, *The Man Who Gave His Company Away*
     (Heinemann, London, 1978).

59   The model rules were approved by the Registrar of Friendly
     Societies and enable associate members of ICOM to register
     under the Industrial and Provident Societies Act 1965, at a
     reduced registration fee.

60   An excellent appraisal of the various legal structures for co-
     operatives and common ownerships, and guidelines for run-
     ning them, is to be found in David H. Wright, *Co-operatives
     and Community* (Bedford Square Press, the National Council
     of Social Service, London, 1979).

61   A revised edition of the excellent S.C.D.C. handbook on
     workers' co-operatives has been published as *Workers' Co-
     operatives – A Handbook* (Aberdeen People's Press, Aber-
     deen, 1980).

62   See Alan Taylor, *Local Co-operative Development Agencies*
     (Co-operative Party, London, 1979).

63 See *In the Making*, 6 for the most comprehensive list of these and other projects, and the I.T.M. supplements in *Undercurrents*. (I.T.M., 44 Albion Road, Sutton, Surrey.)

64 Michael Young and Marianne Rigge, *Mutual Aid in a Selfish Society* (Mutual Aid Press, London, 1979). The authors are respectively Chairman and Director of MAC.

65 *Job Ownership* (JOL, London, 1978).

66 Alastair Campbell et al., *Worker Owners: The Mondragon Achievement* (Anglo-German Foundation, London, 1977).

67 Jenny Thornley, 'Workers' Co-operatives and the Co-operative Development Agency', *Labour Research*, November 1978.

68 The Special Programmes (the Youth Opportunities Programme and the Special Temporary Employment Programme) which started in April 1978 are measures aimed to provide temporary employment, training and work experience for unemployed school leavers and the long-term adult unemployed. They replace the Job Creation Programme, begun in 1975, also administered by the M.S.C.

69 Scottish Development Agency, Welsh Development Agency and the Northern Ireland Development Agency.

70 The H.I.D.B. drew the idea for community co-operatives from the very successful Comharchumann Chois Fharraige enterprises in Gaeltacht, Connemara in Eire. This community co-operative has been running since 1970 and is the largest non-agricultural co-operative in the Republic. It employs over fifty local people full-time in printing, publishing, land reclamation, horticulture and fish farming. Additionally, the co-operative runs Irish language colleges during the summer and maintains a full-time research division to explore new markets and products.

71 Craigmillar Festival Society, *Craigmillar's Comprehensive Plan for Action* (Craigmillar Festival Press, Edinburgh, 1978).

72 Nearly 100 of the staff are sponsored under M.S.C. employment programmes.

73 Craigmillar Festival Society, Edinburgh, 1978.

74 Ibid.

75 Foundation for Alternatives, *Local Initiatives in Britain*, Parts 1 and 2 (The Rookery, Adderbury, Oxford, January and March 1980).

76 Foundation for Alternatives, *Local Enterprise Trusts* (Anglo German Foundation, London, 1978).
77 Bryan Stout, 'Use of Redundant Buildings', *Architects' Journal*, December 6th, 1978.

5 Alternative Organizations in the U.S.A.

1 Richard Gardner, *Alternative America* (Box 134 Harvard Square, Cambridge, Massachusetts 02138).
2 Social Science Institute, Harborside, Maine, 1954.
3 Schocken Books, New York, 1970.
4 The Center now publishes a twenty-page bi-monthly news-paper, *A.T. Times*, excellently produced and very informative.
5 For addresses and details of these and other small farm organizations in the U.S.A., see TRANET's *Directory of A.T. Centers*, 1978; and *Appropriate Technology – a Directory of Activities and Projects*, National Science Foundation, 1977.
6 More accurately, *one* of the energy-saving cities in the U.S.A. Others include Seattle (Washington), Hartford (Connecticut), Northglen (Colorado), Clayton (New Mexico), Ames (Iowa) and Greensboro (North Carolina). See James Ridgeway, *Energy-Efficient Community Planning* (J. G. Press, Emmaus, Pennsylvania, 1979).
7 For details of the Davis programme, see *The Davis Experiment* (published by The Elements, 1747 Connecticut Avenue, Washington D.C., 20009, 1977).
8 A report prepared by a U.S.D.A. Study Team on Dynamic Farming, United States Department of Agriculture, 1980 (U.S. Government Printing Office, Washington, D.C., 1980).
9 *Global 2000, Report to the President*. Vol. 1 is the summary, Vol. 2 the technical report and Vol. 3 the technical documentation on the global models (U.S. Government Printing Office, Washington, D.C., 1980).

6 Canadian Initiatives

1 The Council was a non-government advisory body to the Minister for Regional Economic Expansion. Unfortunately for

rural Canada, it was disbanded by the Minister early in 1979.

2 See 'A Proposed Strategy for the Development of Northern Manitoba', March 1976 (Department of Northern Affairs, Government of Manitoba).

3 Available from Printing and Publishing, Supply & Services, Canada, Ottawa, Canada, K1A OS9.

4 Ibid., p. 91.

5 Thomas Berger, *Northern Frontier, Northern Homeland, The Report of the Mackenzie Valley Pipeline Inquiry*, Vol. 1 (Printing and Publishing, Supply & Services, Canada, Ottawa, 1977).

6 Rein Peterson, *Small Business: Building a Balanced Economy* (Press Porcepic, Ontario, 1977).

7 William A. Dyson, *The Nature of the Economy in a Familial Society*. Report of a seminar held on October 12th–13th, 1977 (Vanier Institute of the Family, 151 Slater, Ottawa).

8 Dr Kenneth T. MacKay, 'Exploration of Self-Sufficiency at the P.E.I. Ark', address to the International Federation of Organic Agricultural Movements, October 1978.

9 Ibid.

7 Economics as if People Mattered

1 James Robertson, *The Sane Alternative* (River Basin Publishing Co., St Paul, Minnesota, 1979).

Supplement A: A.T. in Developing Countries

1 Other institutes of technology, notably those of Delhi and Madras, are also becoming involved in A.T.

2 J. P. Narayan, the great champion of India's poor, was among those who persuaded Schumacher to visit India in the early 1960s. He and Schumacher became close friends.

3 I.T.D.G. internal report, 1977.

4 Each would have the effect of reducing the required size of plant. Thus if the cycle could be cut from its present sixty days to twenty the plant could be about one-third of its present size and cost; and it could be halved again if the gas output per lb. of cow dung were then to be doubled.

5 This change of emphasis has come about because, for instance, they found that the excellent hospital at Gandhigram headquarters was dealing mostly with cases of malnutrition, and that their school was educating young people who had to leave their community in order to earn a living.

6 I.T.D.G. carried out an initial feasibility study and outline proposal for the establishment of such a Centre at Kumasi. Thanks to the help of the Inter-Universities Council, the detailed work of getting the Centre operational was done by H. Dickinson, Department of Electrical Engineering, University of Edinburgh.

7 In 1977/8 the Centre's total income was:

	Cedis ($\mathcal{C}$)
University subvention	82,794
Overseas grants	190,551
Consultancy fees	7,512
Sales by production units	216,451
	497,308

8 *The Diffusion of Intermediate Technology in Ghana: the Work of the T.C.C.*, report by Sally Holterman, an economist commissioned to make this study by Intermediate Technology Industrial Services Ltd, in 1978.

9 From *How to Start a Village Polytechnic*, a manual issued by the Government of Kenya.

10 The Relevant Technology Project is part of a programme of integrated education in Plateau State. For fuller information about it see *Bernard van Leer Foundation Newsletter*, Nos 15 (March 1975) and 23 (Spring 1978).

11 George Macpherson and Dudley Jackson, 'Village Technology for Rural Development: Agricultural Innovation in Tanzania', *International Labour Review*, February 1975.

12 Ibid.

13 The head of this project, George Macpherson, subsequently published an excellent account of this work in *First Steps in Village Mechanization*, Tanzania Publishing House, Dar Es Salaam, 1975.

14 For a detailed appraisal of this programme, see H. Dickinson and Jan Neggers, *Report of a Short Mission to the Centro*

Desarrollo Integrado Las Gaviotas (University of Edinburgh, 1978).

15 There are now two organizations pursuing this kind of approach. The new one is Comunidad por Los Niños. Its director, who built up Futuro, is Alvaro Villa.

Supplement B: Further Alternative Activities in Britain

1 For example, see C. C. Pocock, Chairman of Shell (U.K.), 'More jobs: A small cure for a big problem', the Ashridge Lecture, 1977 (Ashridge Management College).

2 John Pearce, *Can We Make Jobs?* (published for the Local Government Research Unit, Paisley College of Technology, by Heatherban Press, Glasgow, 1978).

3 *An Alternative Energy Strategy for the United Kingdom*, available from CAT.

4 *Undercurrents* 28, June/July 1978.

5 John Jeffery et al., 'Appropriate Technology and Possible Futures', in *Alternative Technology and Institutional Change*, Conference Papers, Department of Applied Science, University of Newcastle, 1977.

6 *Engineering*, Education Supplement, June 1977.

7 All published by the Open University Press, Milton Keynes.

8 Peter Read, *Alternative Technology, Self-Sufficiency and the Future of Work*, A.T.G. Paper No. 6, Open University, 1979.

9 See Robert and Brenda Vale, *The Autonomous House* (Thames and Hudson, London, 1975. Universe Books, New York, 1976); Robert Vale, *Smallholdings and Food Production—An Alternative Strategy for U.K. Agriculture*, A.T.G. Paper No. 1 (Open University, Milton Keynes, 1977); Robert and Brenda Vale, *The Self-Sufficient House* (Macmillan, London, 1980).

10 *Local Initiatives in Great Britain*, Parts 1 and 2 (The Foundation for Alternatives, The Rookery, Adderbury, Oxfordshire, OX17 3NA, 1980).

11 Published in IFDA Dossier No. 15, January/February 1980, 2 Place du Marché, CH 1260 Nyon, Switzerland.

12 Contractors and building firms are beginning to include similar fixtures in speculative private housing development –

in Blythe and Croydon, for example. The scope for future use of renewable energy sources has encouraged a proliferation in the number of small firms now dealing in solar and wind power technologies; there are now over 200 firms in the U.K. marketing solar collecting devices. In 1978 alone more than 4,000 solar collecting devices were installed for office, housing and recreational use in the U.K.

13   H. Liddell and D. Hodges, *Low Energy Housing: Bransholme, Hull*, Alternative Technology Group, Hull College of Higher Education. See also *Undercurrents* 32, 1979.

14   See Howard Liddell (Ed.), *The Aberfeldy Project: A Study in Self-sufficiency* (Hull College of Higher Education, published by Polygon Press, 1979).

15   A recent estimate has put the number of non-member, participating helpers in all local F.O.E. groups at 20,000.

16   Penguin Specials, Harmondsworth, 1972. Houghton Mifflin, Boston, 1972.

17   *Prospectus for a Third Garden City*, T.C.P.A., 17 Carlton House Terrace, London SW1Y 5AS.

18   Konrad Smigielski quoted in Alan Jones, 'The Stanford Plan', in *Natural Energy and Living* No. 4, 1978.

19   See Colin Pritchard, *Ecology, Equity and Ethics* (1977), and same author, *From Here to Where?* (1978); John Carrie, *The Iona Solar Energy Project* (1978); and also *Renewable Energy Sources in Highland Communities* (1978), all published by the Church of Scotland, 121 George Street, Edinburgh.

20   River Basin Publishing Co., St Paul, Minnesota, 1979.

21   An edited collection of reprints of articles from *A.T./U.K. Exchange* and elsewhere has been published: John Davis, *Technology for a Changing World* (I.T. Publications Ltd, 1978).

22   Schocken Books, New York, 1977.

23   See, for example, *Undercurrents* No. 41 (1980), which is largely devoted to co-operatives and includes a detailed guide to recent literature on co-operatives and worker-ownership.

ORGANIZATIONS CONCERNED WITH APPROPRIATE TECHNOLOGY

The addresses of the principal organizations, whose work is referred to in this book, are listed below. For a more comprehensive coverage, the reader is referred to the *Appropriate Technology Directory* (O.E.C.D. Development Centre, 2 rue André Pascal, 75775 Paris Cedex 16); *Appropriate Technology – A directory of activities and projects* (U.S. Department of Commerce, Springfield, Virginia 22161, U.S.A.); and the Intermediate Technology Development Group's *List of Organizations and Centres Concerned with Alternatives* in agriculture, education, energy, health and technology in the U.K. News about existing organizations, and about new ones, is featured in *In the Making* (44 Albion Road, Sutton, Surrey), the bulletin of the Futures Study Centre (15 Kelso Road, Leeds LS2 9PR), and James Robertson's *Turning Point* (10 New Road, Ironbridge, Telford, Salop TF8 7AU).

Part One

Appropriate Health Resources and Technologies Action Group
   (AHRTAG),
85 Marylebone High Street,
London W1M 3DE

Appropriate Technology Development Association,
P.O. Box 311,
Ghandi Bhawan,
Lucknow – 226001,
Uttar Pradesh,
India

Appropriate Technology International,
1709 N Street NW,
Washington, D.C. 20036,
U.S.A.

Arusha Appropriate Technology Project,
P.O. Box 764,
Arusha,
Tanzania

Brace Research Institute,
McDonald College,
McGill University,
Ste. Anne de Bellevue,
Quebec H0A 1CO,
Canada

Centro Mesoamericano de Estudios Sobre Tecnologia Apropriada
    (CEMAT),
8a Calle 6–06, Zona 1,
Edificio Elma, Apto, 602,
Apartado Postal 1160,
Guatemala City,
Guatemala

Commonwealth Secretariat,
Marlborough House,
Pall Mall,
London SW1

Federal Ministry for Economic Co-operation,
Bonn,
Federal Republic of Germany

Friends of the Earth (F.O.E.),
9 Poland Street,
London W1V 3DG

Friends of the Earth (F.O.E.)
124 Spear Street,
San Francisco, Ca. 94105

Indian Planning Commission,
Yojana Bhavan,
Parliament Street,
New Delhi 110001,
India

Intermediate Technology Industrial Services,
Myson House,
Station Road,
Rugby CV21 3HT

International Institute for Environment and Development,
10 Percy Street,
London W1

International Institute for Tropical Agriculture,
P.M.B. 5320,
Ibadan,
Nigeria

International Labour Office (I.L.O.),
C.H. – 1211 Geneva 22,
Switzerland

Job Ownership Ltd,
42 Hanway Street,
London W1

Ministry of Overseas Development (Overseas Development
    Administration),
Eland House,
Stag Place,
London SW1E 5DH

Organization of the Rural Poor,
Ghazipur,
Uttar Pradesh,
India

Société Africaine d'Etudes et de Développement,
Boite Postale 593,
Ouagadougou,
Upper Volta

South Pacific Appropriate Technology Foundation,
Office of Village Development,
P.O. Box 6973,
Boroko,
Papua New Guinea

Technische Ontwikkeling Ontwikkelingslanden (TOOL),
Maurittskade 61A,
Amsterdam,
The Netherlands

Transnational Network for Appropriate Technology (TRANET),
Box 567,
Rangeley,
Maine 04970,
U.S.A.

United Nations Children's Fund (UNICEF),
866 United Nations Plaza,
New York,
N.Y. 10017,
U.S.A.

Volunteers in Technical Assistance (VITA),
3706 Rhode Island Avenue,
Mt Rainier,
Maryland 20822,
U.S.A.

World Health Organization,
20 Avenue Appia,
1211 Geneva,
Switzerland

The Worldwatch Institute,
1776 Massachusetts Avenue NW,
Washington, D.C. 20036,
U.S.A.

Part Two

California State Office of Appropriate Technology,
1530 10th Street,
Sacramento,
California 95814,
U.S.A.

The Canadian Federation of Independent Businesses,
15 Coldwater Road,
Don Mills,
M3B 3JI,
Canada

Centre for Alternative Industrial and Technological Systems
(CAITS),
North East London Polytechnic,
Longbridge Road,
Dagenham,
Essex RM8 2AS

Centre for Community Economic Development, and Institute for
Community Economics,
639 Massachusetts Avenue,
Cambridge,
Massachusetts 02139,
U.S.A.

Centre for the Integration of Applied Sciences,
1308 Acton Street,
Berkeley,
California 94706,
U.S.A.

Centre for Neighbourhood Technology,
570 West Randolph,
Chicago,
Illinois 60602,
U.S.A.

Centre for Rural Affairs, and Small Farm Energy Project,
P.O. Box 405,
Walthill,
Nebraska 68067,
U.S.A.

Combined Organic Movement for Education and Training
(COMET),
Well Hall,
nr. Alford,
Lincolnshire

Control Data Corporation Agricultural Development Centre,
8100 34th Avenue South,
P.O. Box O,
Minneapolis,
Minnesota 55440,
U.S.A.

Co-operative Development Agency,
20 Albert Embankment,
London SE1 7TJ

Council for Small Industries in Rural Areas,
Queen's House,
Fish Row,
Salisbury,
Wilts SP1 1EX

Cornerstones,
54 Cumberland Street,
Brunswick,
Maine 04011,
U.S.A.

Country College,
Well Hall,
Nr. Alford,
Lincolnshire

Craigmillar Festival Society,
108 Mountcastle Drive South,
Edinburgh EH15 3LL

Ecology Action, and Common Ground,
2225 El Camino Real,
Palo Alto,
California 94306,
U.S.A.

Emerson College,
Forest Row,
Sussex RH18 5JH

Farallones Institute,
The Rural Centre,
15290 Coleman Valley Road,
Occidental,
California 95465,
U.S.A.

and

Integral Urban House,
1516 5th Street,
Berkeley,
California 94710,
U.S.A.

Federation of Working Communities,
5 Dryden Street,
London WC2

The Foundation for Alternatives,
The Rookery,
Adderbury,
Nr. Banbury,
Oxfordshire OX17 3NA

Henry Doubleday Research Association,
Convent Lane,
Bocking,
Braintree,
Essex

Highlands and Islands Development Board,
Community Co-operatives Programme,
16 South Beach,
Stornoway,
Isle of Lewis

Home Workers Organized for More Employment (HOME),
Orland,
Maine 04472,
U.S.A.

Industrial Common Ownership Finance Ltd (ICOF),
1 Gold Street,
Northampton NN1 1SA

Industrial Common Ownership Movement (ICOM),
Beechwood College,
Elmete Lane,
Roundhay,
Leeds LS8 2LQ

The Institute for Local Self-Reliance,
1717 18th Street NW,
Washington, D.C. 20009,
U.S.A.

Institute of Man and Resources,
Little Pond,
Souris, R.R. No. 4,
Prince Edward Island,
C0A 2BO,
Canada

Inter-Action,
15 Wilkin Street,
London NW5

Intermediate Technology Development Group,
9 King Street,
London WC2E 8HN

Inwork,
c/o John Morrison,
Fife Regional Council,
Education Department,
15 Greensfield,
Kirkcaldy,
Fife

Mutual Aid Centre,
18 Victoria Park Square,
London E2 9PF

National Centre for Appropriate Technology,
P.O. Box 3838,
Butte,
Montana 59701,
U.S.A.

New Alchemy Institute,
Box 47,
Woods Hole,
Massachusetts 02543,
U.S.A.

New Mexico Solar Energy Association,
P.O. Box 2004,
Santa Fe,
New Mexico 87501,
U.S.A.

New School for Democratic Management,
580 Howard Street,
San Francisco,
California 94105,
U.S.A.

Operation Brotherhood,
3475 West Ogden Avenue,
Chicago,
Illinois 60623,
U.S.A.

Organic Farmers and Growers Co-operative (also International
    Institute for Biological Husbandry),
9 Station Approach,
Needham Market,
Ipswich,
Suffolk

People's Development Corporation,
500 E 167th Street,
Bronx,
New York, N.Y. 10456,
U.S.A.

The Pye Research Institute,
Walnut Tree Manor,
Haughley Green,
Stowmarket,
Suffolk

Shelter Institute,
38 Center Street,
Bath,
Maine 04630,
U.S.A.

Soil Association,
Walnut Tree Manor,
Haughley Green,
Stowmarket,
Suffolk

Sudbury 2001,
P.O. Box 1313,
Sudbury,
Ontario,
P3E 4S7,
Canada

The Vanier Institute of the Family,
151 Slater Street,
Ottawa,
Ontario,
K1P 5H3,
Canada

Supplement A

Allahabad Polytechnic,
Allahabad 211002,
U.P. State,
India

Application of Science and Technology in Rural Areas (ASTRA),
Indian Institute of Science,
Bangalore 560012,
India

Appropriate Technology Development Organization,
House No. 1-B,
Street 47th, F-7/1,
Islamabad,
Pakistan

Birla Institute of Technology,
SIRTDO,
Ranchi,
Bihar State,
India

Indian Institute of Technology (Bombay),
Powai,
Bombay 400076,
India

Khadi and Village Industries Commission,
'Gramodaya',
3 Irla Road, Ville Parle (West),
Bombay 400056,
India

Planning Research and Action Department of the U.P. State
    Government,
Lucknow,
U.P. State,
India

Sarvodaya Appropriate Technology Development Programme,
Lanka Jatika Sarvodaya Shramadana Sangamaya (Inc.),
77 De Soysa Road,
Moratuwa,
Sri Lanka

SENA,
Subdirección Tecnologico Pedagogico,
Apartado Aereo 9801,
Bogotá,
Colombia

Tanzania Agricultural Machinery Testing Unit (TAMTU),
P.O. Box 1389,
Arusha,
Tanzania

Technology Consultancy Centre (T.C.C.),
University of Science and Technology,
University Post Office,
Kumasi,
Ghana

Village Technology Unit (Kenya),
c/o UNICEF,
Box 44145,
Nairobi,
Kenya
and
c/o Youth Development Programme (Training),
Department of Social Services,
P.O. Box 30276,
Nairobi,
Kenya

Supplement B

The Achloa Centre,
Aberfeldy,
Perthshire,
Scotland

The Centre for Alternative Technology (CAT),
Llwyngwen Quarry,
Machynlleth,
Powys,
Wales

Ecology Party,
121 Selly Park Road,
Birmingham

Green Alliance,
16 Strutton Ground,
London SW1

Natural Energy Centre,
2 York Street,
London W1H 1FA

Network for Alternative Technology and Technology Assessment
    (NATTA),
c/o Faculty of Technology,
Open University,
Walton Hall,
Milton Keynes MK7 6AA

New Age Access,
24 Fore Street,
Hexham,
Northumberland

Open University,
Walton Hall,
Milton Keynes MK7 6AA

Parliamentary Liaison Group for Alternative Energy Strategies
   (PARLIGAES),
Flat 2,
81 Onslow Square,
London SW17 3LT

Society, Religion and Technology Project,
Church of Scotland,
121 George Street,
Edinburgh EH2 47N

FURTHER READING ON APPROPRIATE TECHNOLOGIES FOR DEVELOPING COUNTRIES (ANNOTATED)

BHALLA, A. S. (Ed.), *Towards Global Action for Appropriate Technology*, Oxford, Pergamon, 1979.
This is a collection of papers prepared by a well known group of experts on appropriate technology for discussion at a meeting convened in December 1977 by the Netherlands government and I.L.O. It is a book for those who are already well versed in the subject, and interested more in the practical implementation of appropriate technologies than in the promotion of concepts.

The paper by Jéquier is particularly refreshing and offers some new insights into the theoretical and political considerations of the subject. He argues that most of the criteria for the adoption of appropriate technologies have been economic and, while these are important, there are a number of social, cultural and technical factors which deserve at least equal attention.

The other papers look at appropriate technology in the light of: a basic needs strategy; the structure, operation and problems of national and regional technology groups and institutions; the continuing activities of U.N. organizations engaged in appropriate technology; and the issue of international mechanisms for the promotion of appropriate technology.

CARR, M. N., *Economically Appropriate Technologies for Developing Countries*, I.T. Publications Ltd, London, 1976.
This is a publication of interest to those seeking evidence that small-scale technologies in the areas of agriculture, manufacturing

industries, housing and infrastructure can be as efficient, or even more efficient, in economic terms, than large-scale technologies.

CONGDON, R. J. (Ed.), *Lectures on socially appropriate technology*, Technische Hogeschool, Eindhoven, Netherlands, 1975.
A collection of twelve lectures delivered as part of a course to provide an introduction to socially appropriate technology. Covers many of the important areas of interest to those who are concerned with the role of appropriate technology in rural development.

DIWAN, R. K. and LIVINGSTON, D., *Alternative Development Strategies and Appropriate Technology: Science Policy for an Equitable World Order*, Pergamon, New York, 1979.
A refreshing and well written book for those development practitioners who are already convinced of the wisdom of development which tackles the problem of poverty at its source by generating productive employment opportunities in rural areas; by raising the productivity and incomes of those already employed in these areas; and by allocating more resources to the provision of basic rural services such as water, power, health, transport and education. It deals with the problems of the practical implementation of such development and raises many questions which have so far been unasked.

DUNN, P. D., *Appropriate Technology: Technology with a Human Face*, Macmillan, London, 1978.
This is a book aimed at the general reader, who should find it both easy and interesting to read.
   The first part of the book covers the growth of the appropriate technology movement and its meaning and purpose within the context of the problems and development objectives of developing countries. It then looks at appropriate technology in practice, describing what technologies are available and giving examples of their application. The areas covered are food, agriculture and agricultural engineering; water and health; energy; services – medicine, transport, building; small industries in rural areas; and education, training, research and development. The large number of diagrams and photographs are very useful for the newcomer to the subject.
   Although much of the work described relates to projects in-

itiated by A.T. institutions in the developed countries, there is also very fair coverage of the work being done by individuals and institutions in the developing countries.

Perhaps the best point of this book is that it relates the technologies it describes to the people who will be using them and to the conditions in which they live and work.

EVANS, D. D. and ADLER, L. N. (Eds), *Appropriate Technology for Development: A Discussion and Case Histories*, Westview Press, Boulder, Colorado, 1979.

This analysis of appropriate technology first explores the concept of development in terms of need, characteristics and theories and then examines the pivotal role of technology in the development process. The twenty case studies in the book, which range from fish preserving techniques in Indonesia to micro-hydro projects in Papua New Guinea, are of varying interest and value. They do, however, add to the still neglected area of the sociological, cultural and economic problems involved in introducing new technologies in rural areas of developing countries.

FRENCH, D., *Appropriate technology in social context*, VITA, Washington, 1977.

Discusses the barriers which exist to the widespread adoption of appropriate technologies in developing countries. Devotes special attention to the need for local participation in decision making.

HOLTERMANN, S., *Intermediate Technology in Ghana: The Experience of Kumasi University's Technology Consultancy Centre*, ITIS, Rugby, 1979.

A useful publication for those interested in the details of the work being carried out by a local Appropriate Technology Centre in a developing country. Besides a chapter on the economic environment in Ghana and the implications of this for the development and dissemination of small-scale technologies, there are detailed sections on the history of the Technology Consultancy Centre and its current programmes. A valuable account is given of how the T.C.C. reacted to, and helped small entrepreneurs to cope with, changes in the economic environment.

JÉQUIER, N., *Appropriate Technology: Problems and Promises*, O.E.C.D., Paris, 1976.

Although written in 1976, this book is still extremely valuable for those seeking to understand the concept and reality of technologies appropriate to people in developing countries. The importance of the 'software' side of appropriate technology is emphasized. The case studies, on which Jéquier's excellent introduction is based, give an insight into the successes and problems involved in the implementation of appropriate technology.

JENKINS, G., *Nonagricultural choice of technique*: *an annotated bibliography of empirical studies*, Institute of Commonwealth Studies, Oxford, 1975.
Useful for those who are interested in showing the wide range of techniques that can be economically efficient in manufacturing industries in developing countries.

OWENS, E. and SHAW, P., *Development Reconsidered*, D. C. Heath, Massachusetts, 1972.
One of the first, and still one of the best books advocating a radically different approach to aid and development, based on appropriate technology. Its theme is that the small peasant farmer and rural artisan are the keys to self-generating development, and the book cites many practical examples of what can be achieved when peasant farmers and small communities are given appropriate technological and organizational support.

RAMESH, J. and WEISS, C. (Eds), *Mobilizing Technology for World Development*, Praeger, New York, 1979.
This book aims at clarifying new development perspectives from which technological choices need to be viewed if the world's industrialized North and its poorer South hope to mobilize technology to achieve sustained growth and increased justice and stability, at home and internationally. Contains the complete report of a symposium held on the subject of mobilizing technology for world development (Jamaica, 1979) and eighteen thought-provoking papers written by internationally distinguished experts on the subject of technology transfer and the implications and practicalities of technological independence in developing countries.

ROBINSON, A. (Ed.), *Appropriate Technologies for Third World Development*, Macmillan, London, 1979.
This contains the papers and proceedings of a conference held by the International Economic Association in 1977 at Teheran, which

examined why the technologies appropriate to circumstances in developing countries are not used. Many of the papers look at the experiences of specific countries and will be particularly useful to those interested in economic and political environments suitable to the widespread adoption of appropriate technologies in developing countries.

STEWART, F., *Technology and Underdevelopment*, Macmillan, London, 1978.
This is a book mainly for economists concerned with the role of industrialization in developing countries. It examines at length the conventional economic thought with regard to choice of technique, employment, dependency and trade. The consequences of distribution of Western technologies are fully explored, with the resultant neglect of rural areas, where people have minimal purchasing power and therefore little ability to affect the market. The advantages of expanding trade between underdeveloped countries are outlined, in addition to the need to search further for low-cost ways of meeting indigenous basic needs.

Of particular interest are the two chapters devoted to empirical studies – one on maize grinding in Kenya and the other on concrete block making in Kenya. Both show clearly how the choice of products affects the choice of techniques and indicate the extent of the influence of consumer preferences on technological choice.

TIMMER, P., et al., *The Choice of Technology in Developing Countries*: *Some Cautionary Tales*, Harvard University, 1975.
Discusses how the prevailing economic and political climate leads to technological choices which adversely affect the rural and urban poor in developing countries. The four detailed case studies in the book cover the choice of technique in: rice milling in Indonesia; irrigation tubewells in East Pakistan; six manufacturing industries in Indonesia; and the petrochemical industry in Colombia. All are classic cases of the choice of 'inappropriate technology'.

UNIDO, *Monographs on Appropriate Industrial Technology*, Vols 1 to 6, United Nations, New York, 1979.
This series of monographs is based on documents prepared for the UNIDO Forum on Appropriate Industrial Technology held in New Delhi in November 1978. Vol. 1 covers the conceptual and policy framework for appropriate industrial technology. The papers in

this volume by Tinbergen, Ranis, Frost, Desai and Chebbi make particularly good reading for those who are concerned with the implementation of policies to help the rural and urban poor in developing countries. This first volume also includes a report of the ministerial level meeting which met in November 1978 in Anand.

For those who are interested in specialized aspects of appropriate technology, the remaining volumes include articles which give an excellent summary of: current developments in low-cost transport (Vol. 2); paper and small pulp mills (Vol. 3); agricultural machinery and implements (Vol. 4); energy for rural requirements (Vol. 5); and textiles (Vol. 6). Of particular interest are the papers by Barwell and Howe (Vol. 2); Mitra (Vol. 4); and Turner (Vol. 5).

A further seven volumes are to be published covering the areas of: food storage and processing; sugar; oils and fats; drugs and pharmaceuticals; light industries and rural workshops; construction and building materials; and basic industries.

FURTHER READING ON APPROPRIATE TECHNOLOGIES FOR THE U.K.

BALDWIN, J. and BRAND, S. (Eds), *Soft-Tech*, Penguin, Harmondsworth, 1978.

BOOKCHIN, MURRAY, *Post-Scarcity Anarchism*, Wildwood House, London, 1974.

BOYLE, G., ELLIOTT, D. and ROY, R. (Eds), *The Politics of Technology*, Longman, Harlow, 1977.

BOYLE, G. and ELLIOTT, D. (Eds), *Radical Technology*, Wildwood House, London, 1976.

BRAY, J. and FALK, N., *Towards a Worker Managed Economy*, Tract No. 430, Fabian Society, London, 1974.

CAMPBELL, A., *Worker Ownership*, Industrial Common Ownership Movement, Leeds, 1976.

C.I.S. REPORT, *The New Technology*, Counter Information Services, London, 1979.

CLARKE, R., *Technological Self-Sufficiency*, Faber and Faber, London, 1976.

COATES, K. and TOPHAM, T., *Workers' Control*, Panther Books, St Albans, 1968.

COCKERTON, P., GILMOUR-WHITE, J., PEARCE, J., and WHYATT, A., *Workers' Co-operatives – A Handbook*, Aberdeen People's Press, 163 King Street, Aberdeen, 1980.

COOLEY, M., *Architect or Bee?*, Langley Technical Services, c/o 95 Sussex Place, Slough SL1 1NN, 1980.

DAVIS, J., *Technology for A Changing World*, I.T. Publications Ltd, London, 1978.

DICKSON, D., *Alternative Technology*, Fontana, London, 1974. Universe Books, New York, 1975.

ELLIOTT, D., *Energy Options and Employment*, CAITS, London, 1979.

ELLIOTT, D. and R., *The Control of Technology*, Wykeham Publications, London, 1976.

FALK, N., *Think Small: Enterprise and the Economy*, Tract No. 453, Fabian Society, London, 1978.

FORMAN, N., *Another Britain*, Bow Publications Ltd, London, 1979.

FOUNDATION FOR ALTERNATIVES, *Local Initiatives in Great Britain*, Part 1, January 1980, Part 2, March 1980 (The Rookery, Adderbury, Oxford).

GEORGE, M., *The Future of Employment in Engineering and Manufacturing*, CAITS, London, 1978.

GERSHUNY, J., *After Post-Industrial Society? The Emerging Self-Service Economy*, Macmillan, London, 1978.

GOYDER, M., *Socialism Tomorrow: Fresh Thinking for the Labour Party*, Fabian Society, London, 1979.

ILLICH, I., *Deschooling Society*, Penguin, Harmondsworth, 1971. Harper & Row, New York, 1971.

——, *Medical Nemesis*, Calder and Boyars, London, 1975. Pantheon, New York, 1975.

——, *Tools for Conviviality*, Calder and Boyars, London, 1973. Harper & Row, New York, 1973.

ILLICH, I., et al., *Disabling Professions*, Marion Boyars, London, 1977.

JONES, D. C., 'The Economics of British Producer Co-operatives', unpublished Ph.D. thesis, Cornell University, U.S.A., 1974.

LEACH, G., et al., *A Low Energy Strategy for the United Kingdom*, International Institute for Environment and Development, Science Reviews, London, 1979.

LOVINS, A. B., *Soft Energy Paths: Towards a Durable Peace*, Penguin, Harmondsworth, 1977. Harper & Row, New York, 1978.

MASON, M., *Creating Your Own Work*, Gresham Books, Old Woking, Surrey, 1980.

MINNS, R. and THORNLEY, J., *Local Government Economic Planning and the Provision of Risk Capital for Small Firms*, Centre for Environmental Studies, 62 Chandos Lane, London WC2 4HH, 1978.

OWEN, D., *Co-operative Ownership*, Industrial Common Ownership Movement and the Co-operative Party, 158 Buckingham Palace Row, London SW1, 1980.

RADICE, G. (Ed.), *Working Power*, Tract No. 431, Fabian Society, London, 1974.

RIFKIN, JEREMY, *Entropy*, Viking Press, New York, 1980.

ROBERTSON, JAMES, *Power, Money and Sex: Towards a New Social Balance*, Marion Boyars, London, 1976.

——, *Profit or People? The New Social Role of Money*, Calder and Boyars, London, 1974.

SANDBACH, FRANCIS, *Environment, Ideology and Policy*, Basil Blackwell, Oxford, 1980.

SEYMOUR, JOHN, *Self-Sufficiency*, Faber and Faber, London, 1976. Schocken, New York, 1976.

STEAD, PETER, *Self-Build Housing Groups and Co-operatives: Ideas in Practice*, Anglo-German Foundation, London, 1980.

TIVEY, L., *The Politics of the Firm*, Martin Robertson, Oxford, 1978.

TURNER, JOHN, *Housing by People*, Marion Boyars, London, 1976.

VALE, R., and B., *The Autonomous House*, Thames and Hudson, London, 1975. Universe Books, New York, 1976.

INDEX

(i) General

Agriculture: Britain, 86–91, 190, 257, 259, 263, 269; Canada, 170, 173–4; Colombia, 243; U.S.A., 48, 129, 132, 136, 144–6, 148–51, 157, 159, 162, 163

Agriculture, chemical: dangers of, 10, 49, 86–7, 147; labour bottlenecks, 47; secondary tasks of, 88; sustainable, 86, 87; *see also* Biological husbandry

Agricultural equipment: Ghana, 228; Kenya, 66; Malawi, 66; Mauritius, 66; Nigeria, 48, 69, 237; Pakistan, 217; Sri Lanka, 221; Swaziland, 66; Tanzania, 66, 239; Zambia, 47–8

Agricultural processing, 61–2, 67, 69, 121, 135, 139, 146, 157, 186, 195, 196, 200, 216, 217, 223, 225, 226; *see also* Food processing

Alternative Sources of Energy, 139–40

Alternative Technology Directory, 280

Appropriate Technology: policy for developing countries, 183–8; policy for industrialized countries, 188–91; service bases, 45, 48, 51, 196, 199, 201, 205–6, 216, 229; *see also* Intermediate technology

Appropriate Technology, 279

A.T. Index, 279

A.T./U.K. Exchange, 278

Biogas energy, *see* Energy

Biological husbandry, 86–90, 132, 140, 145–8, 152, 163; *see also* International Institute of Biological Husbandry

Capital, declining return on, 83

Capital intensive industry, destructive effect of, 23, 76–80, 83, 167, 169, 179, 203, 207

Capital saving, 29, 200, 205, 211, 242; *see also* Cost per workplace

City growth, 9–10, 23, 26, 33

Community development corporations, 133, 134

Community initiatives, 117–26, 129, 135–8, 142, 143, 154, 155, 159, 161–2, 261–6; management training for, 110, 153

Conserver society, 165–7, 189–91

Construction and building materials: Botswana, 43; Britain, 44, 265, 274–5; Canada, 174; China, 210; Colombia, 245; Egypt, 42; Gambia, 42, 43; Ghana, 42; Honduras, 69; India, 41, 71, 194, 196, 198, 210–11; Kenya, 40–1; Nigeria, 40; Pakistan, 218; South Sudan, 42; Sri Lanka, 43; Tanzania, 41, 42; U.S.A., 130, 131, 137, 138, 145

Co-operatives: Britain, 88, 98, 103–17,

(ii) Organizations

(iii) Universities and Colleges